The Kaiparowits Coal Project and the Environment

A CASE STUDY

by
Ronald Jepperson
Eric Hemel
Robert Hauptman

Stanford Western Energy Policy Study
Graduate School of Business
Stanford University
Public Management Program

Ronald E. Wyzga
EPRI Project Manager

ANN ARBOR SCIENCE
PUBLISHERS INC / THE BUTTERWORTH GROUP
in cooperation with
ELECTRIC POWER RESEARCH INSTITUTE
EPRI EA-1620

Published 1981 by Ann Arbor Science Publishers, Inc.
P.O. Box 1425, 230 Collingwood, Ann Arbor, Michigan 48106

Library of Congress Catalog Card Number 80-53420
ISBN 0-250-40399-4

Notice:

This report is an account of work prepared by the Stanford Western Energy Policy Study, Graduate School of Business, Stanford University, Public Management Program, and sponsored by the Electric Power Research Institute, Inc. (EPRI). Neither the contractor, EPRI, members of EPRI or any person acting on their behalf: (a) makes any warranty or representation, express or implied, with respect to the accuracy, completeness or usefulness of the information contained in this report, or that the use of any information, apparatus, method or process disclosed in this report may not infringe privately owned rights; or (b) assumes any liabilities with respect to the use of, or for damages resulting from the use of, any information, apparatus, method or process disclosed in this report.

Keywords: Coal-fired Power Plant; Kaiparowits; Environmental Impacts; Energy Needs

Foreword

Originally, the Kaiparowits project seemed to provide a promising solution to environmental and economic concerns. It was to be sited in a remote location, far from population centers which it might otherwise affect, and near an abundant source of low-sulfur coal. Before the project was terminated, it became a rallying point for environmental opposition, which characterized the proposed project as "the ultimate obscenity." The divergence of opinion and changes in outlook over time make Kaiparowits an interesting object of study, particularly since many similar projects are foreseen for the western United States.

This study focuses on the role that environmental politics played in cancellation of the Kaiparowits project. The history and resolution of the Kaiparowits controversy should provide useful analytical insights into the important issues of energy economics and environment and public policy. These issues must be resolved if energy resources in the western United States are to be developed.

Ronald E. Wyzga
Electric Power Research Institute

Acknowledgment

This study was prepared by the Stanford Western Energy Policy Study for the Electric Power Research Institute. Ronald Jepperson was the principal author. Eric Hemel directed the study project and was a contributing author.

Robert Hauptman did much of the research work for Chapter 6, and played a principal role in authoring the chapter. Lynne Parshall contributed an initial draft of Chapter 2. Mark McConnell and Andrew Schwartz conducted research for the project. The authors are thankful to Sharon Aglito, who provided help, guidance, and editorial assistance at every stage of the project. Thanks also to Professor Henry Rowen and Thomas Neville, who reviewed and commented on drafts of the study.

Preparation of this study involved a number of personal interviews with utility officials, government officials, and environmentalists. We are especially grateful to officials of Southern California Edison and officials of the United States Department of the Interior. The authors, however, accept full responsibility for the study's interpretations and conclusions.

Contents

1
Introduction

On April 14, 1976, after 13 years of planning, Southern California Edison announced that it would no longer participate in the construction of the Kaiparowits power plant in southern Utah. Kaiparowits was to have been the largest coal-burning plant in the United States, consuming nine million tons of low-sulfur coal a year and producing 3,000 megawatts of electricity—enough to supply a city of three to four million people. The project was to encompass the generating station, with its four 600-foot stacks; four underground coal mines; a limestone quarry; a new town for as many as 15,000 people; and 1460 miles of new electricity transmission lines. It was to provide electricity to urban areas of Southern California and Arizona through its group of owners, Southern California Edison Company, San Diego Gas and Electric Company, Arizona Public Service Company, and Salt River Project (of Arizona).

But the location of the plant site was to significantly affect the project's history. The plant was to be built on the remote Kaiparowits plateau, within 250 miles of eight national parks, 26 national monuments, three national recreation areas, and three national historic sites, including Grand Canyon, Zion, and Bryce Canyon National Parks. An *Audubon Magazine* article, presenting an environmentalist position on the project, described the plateau area as

> one of the least disturbed of all of southern Utah's remaining wild places, literally in the heart of the Southwest's circle of major parks and monuments. A long, formidable, desert-and-rock highland of 1,600 square miles, sloping from the Aquarium Plateau in the north, where it rears almost a mile and a half high, to Lake Powell, where jagged escarpment and canyons descend 3,000 feet to the water, it is a silent world of barren, cliff-bound benches, massive terraced plateaus, vertical rock walls, and deep canyons. Much of the brilliant, eroded landscape, bare of vegetation, glows with colors that range from bright red to umber, yellow, gray, and white [1].

Although the plant was to have had the most sophisticated pollution control equipment available, environmentalists and many government officials believed that the 80,000 or so tons of contaminants that would annually escape to the atmosphere would adversely affect the area's scenic resources, especially in conjunction with operation of the nearby Navajo coal plant (36 miles away) and the more distant Four Corners coal facility near Farmington, New Mexico. Moreover, many argued that the need for the project had never been conclusively demonstrated. By the time of the project's cancellation, Kaiparowits had become a national issue, a symbol of the debate over the United States' energy and environmental objectives.

When the participating utilities began to consider the project in the early 1960s, no one would have expected the long battle that was to ensue. From a utility perspective, Kaiparowits seemed like an ideal way to meet Southern California's increasing electricity needs. Kaiparowits would utilize abundant, low-sulfur coal from deep (as opposed to strip) mines; it was economically attractive; it would not add to pollution problems in Southern California; and it would help southern Utah's underdeveloped economy.

In 1965, David Brower, then president of the Sierra Club, seemed to concur with the utilities' view of Kaiparowits, stating that a coal plant on the Kaiparowits plateau would be preferable to and cheaper than the construction of more hydroelectric dams on the Colorado River.

In January 1964, the project participants took their first step, filing with the state of Utah to obtain water from the Colorado River to be used as coolant at the plant. Utah granted the request, but because the water was impounded in a federal area, Lake Powell, application also had to be made to the federal Bureau of Reclamation. The utilities had to wait until late 1969 for their request to be granted, due to federal-state disagreements over the control of Colorado River water rights. Environmental concern had been growing steadily during the interim years, and environmentalists were becoming more and more concerned about the impacts of siting coal-fired plants in the Southwest. In June 1973, Secretary of the Interior Rogers Morton announced that he would delay decisions on further power plant construction in the Southwest until a regional study assessing the need for and environmental effects of such construction would be completed. (The Department of the Interior has ultimate jurisdiction over any project involving federal lands; large portions of Southwest lands are federally owned.) Later, the completion of a detailed environmental impact statement (EIS), in compliance with the National Environmental Policy Act (NEPA) of 1969, as well as increasingly vocal criticism of the project by environmentalists, further delayed Kaiparowits.

Over the years, what the utilities had projected to be a $500 million project had become a $3.5 billion one due to changes in plant design (for example, pollution abatement equipment was not included in the original design), the inclusion of town planning and coal mine development costs, changes in cost estimation procedures, inflation, and increasing fuel, labor, and equipment costs. Considering the changed economics of the project, taking into account the drop in electricity demand that followed the OPEC oil embargo, and anticipating protracted environmentalist litigation even if the Secretary of the Interior were to approve construction, the utilities terminated the project.

The controversy that led to the demise of Kaiparowits reflected not only differing analyses of the need and effects of the project, but more fundamentally, different perceptions of the relationships between energy expansion, economics, and environmental quality. This case study will analyze the Kaiparowits project history in order to raise a number of important environmental, political, regulatory, and economic issues associated with energy expansion and environmental protection. Chapter 2 will outline the project's history, from the utility decision to site plants outside of California to the Kaiparowits withdrawal. Chapter 3 will characterize the uncertainty over Kaiparowits' environmental effects and the differing interpretations of the Kaiparowits Environmental Impact Statement. This chapter will also present the debate over the need for the Kaiparowits project and the dispute over energy demand projections. Chapter 4 will concentrate on the environmentalist role in the project's history. Chapter 5 will analyze the federal regulatory context and the decision-making processes in the Department of the Interior relevant to Kaiparowits. Chapter 6 will discuss economic and financial aspects of the project in order to characterize the changing constraints on utility decision-making and the different factors underlying the utility decision to withdraw the Kaiparowits application. Chapter 7 will conclude by raising a number of questions regarding the wider effects and implications of the Kaiparowits withdrawal.

The Kaiparowits story should illustrate some of the facets of our present energy and environmental planning dilemma. Utilities must plan to meet demand for electricity in a context where they can anticipate neither the length of the planning and construction period for a project nor the probabilities of the project actually reaching the point of operation. Government is called on to regulate in a context of great uncertainty over the environmental effects of energy expansion and to balance energy "benefits" with environmental "costs" where there is no consensus on how to measure or value environmental integrity or energy benefits. Such trade-offs are inherent in any government energy or environmental policy decision, whether they are made explicitly or not, and any such decision involves implicit commitments to certain energy or environmental policy paths as against others. For the most part, however, regulators proceed on an incremental plant-by-plant basis, overlooking these larger policy contexts. The Kaiparowits history invites the reader to assess this policy dilemma, and to determine for himself the implications of the conflict over Kaiparowits.

2
Historical Overview

Kaiparowits' 13-year history can appear at first glance to be incomprehensibly long, complicated, and costly. Perhaps the Kaiparowits story becomes more clear when considered in the context of the attempt to use Southwest coal for the production of California and Arizona electricity and the even larger context of changing environmental regulation and politics. Kaiparowits project history marked the first time that the issue of the export of California pollution for the import of electricity became a national one. Kaiparowits history was complicated by the number of actors involved, the cast including the participating utilities, the states of Utah, Arizona, and California, local Utahns, environmental groups, Senators and Congressmen, the Department of the Interior (DOI), and the Environmental Protection Agency (EPA). In addition, Kaiparowits history was prolonged by regulatory conflict and change, the changing climate of opinion regarding the environment, and radically changing energy economies. This chapter will present an outline of the most important events in the Kaiparowits chronology, in order to prepare the way for more detailed examination in later chapters.

The formal history of the Kaiparowits power project began on June 18, 1963, when the resource subsidiaries of Arizona Public Service, San Diego Gas and Electric and Southern California Edison formulated the Kaiparowits Coal Project Agreement. To be understood, however, the Kaiparowits proposal must be situated within the history of the attempts by Southern California's utilities to build generation capacity outside of California. Stringent environmental controls were developed earlier in California than in most other states. These controls were unusual, but then so were the environmental problems that California faced. The hot, dry, windless weather from spring through fall results in temperature inversions. These inversions, which trap polluted air, regularly combine with the sunshine to produce the photochemical reactions resulting in smog. Serious pollution problems in California first began in the 12,000-square mile Los Angeles air basin. The first major smog attack was in September of 1943, and after WW II ended, local officials began a campaign to combat the smog.

In 1947, the Los Angeles County Air Pollution Control District (APCD) was formed to deal with these pollution problems. It implemented its decisions with a permit system, requiring permits for any modification of existing polluters and for all new polluters. At this time there was no knowledge of the photochemical effect that caused the smog, so the APCD at first concentrated on visible polluters. The first major targets of the APCD were the oil refineries and the iron smelting industries, and by 1951 the black smoke and soot problem was under control.

In the mid-1950s, the APCD turned its sights toward generating plants. The APCD had found what appeared to be a relatively simple solution to the generating plant pollution problem. It believed that pollution-free natural gas could replace the high-sulfur oil being burned, thereby eliminating most of the serious pollution effects of power production in Los Angeles County. In 1956, the APCD requested the California Public Utilities Commission (PUC) to seek natural gas supplies for in-state power generation use.

Problems with the APCD push for natural gas usage were suggested as early as 1958, when the Federal Power Commission (FPC) warned that utilities could not continue to rely on the limited supply of natural gas to meet generating needs. The APCD, however, continued to believe that it had discovered an ideal solution to its pollution problem, and by 1964 was requiring that utilities rely entirely on natural gas when available. The amount of natural gas used in the Los Angeles basin increased from 10 million barrels oil equivalent to 40 million barrels during the years from 1955 to 1966. In 1970, the FPC restricted future growth of natural gas supplies by limiting construction of new gas pipelines to California, believing that natural gas was more appropriately allocated to private residential use.

Given that the APCD was not going to allow high-sulfur oil to be burned near the major load centers in California, and also given the tight federal restrictions on the substitute fuels of natural gas and low-sulfur imported fuel oil, it is not surprising that the utilities began to look for other types of energy production in the early 1960s.

In its 1968 prospectus, Southern California Edison (SCE) revealed that it had no plans for future development in California once existing projects were completed. SCE's final effort to construct a conventional plant in California was dropped in 1971, after a California Supreme Court decision allowed the California Air Resources Board to prohibit the building of a plant at Huntington Beach despite approval by the California PUC.

SCE believed that hydroelectric and nuclear generation were the only other options for in-state power production. Hydroelectric generation had obvious limitations and could be counted on for only a small increase in power supply; construction of nuclear plants had always been difficult because of the close supervision of the Atomic Energy Commission (AEC). The AEC found special difficulties with nuclear development in California because of the state's seismological problems. A planned nuclear plant at Bodega Head was dropped by Pacific Gas and Electric (PG&E) in 1960 because of local opposition to the location of transmission lines and increasing concern over safety matters. Although nuclear plants were not eliminated from all utility resource plans, it became apparent that they were not going to be able to handle all of the growing California demand.

Out-of-state development thus looked attractive even in the early 1960s, not only because of the fuel shortages and the California regulatory control, but also because of a conscious desire on the part of SCE officials to avoid the pollution problems that construction in California would involve. The obvious out-of-state options were in Nevada, New Mexico, and Arizona, with Colorado and Wyoming too far away for serious consideration. Vast quantities of coal were available, water was scarce but could be obtained, and environmental regulations were much less stringent in each of these states.

Construction on the first two major out-of-state plants, Four Corners and Mohave, began in the early 1960s.* The Four Corners plant near Farmington, New Mexico, was the first large coal-burning plant in the Southwest. Arizona Public Service, Southern California Edison, and Salt River Project all participated in this project. Southern California utilities also participated in the Mohave plant in Clark County, Nevada. The Four Corners plant was built with the cooperation of the Department of the Interior acting for the Indian tribes that owned a large portion of the land in the area on which the plant was to be built and over which transmission lines would cross. In both cases, Indians controlled large portions of the unused water that was essential for power production.

---

* Four Corners was completed in 1970, Mohave in 1971.

Construction of additional generation units at Four Corners or Mohave faced a number of prohibitive problems. The participating companies were beginning to anticipate difficulties with the Hopi Indians. The Hopis and the Navajos together held the lease on the strip mine that provided Mohave with 100 per cent of its coal, yet they were becoming less and less enamored with Southwest power plant construction. There was no water availability problem at Four Corners, but environmentalists had labeled this plant as the single greatest polluter in the U.S. and would have reacted vehemently to any addition at the site.

SCE planned instead to become a 50 per cent participant in the proposed Navajo project, near Page, Arizona, but negotiations with the Indian lessors were not going well.* A 5000-megawatt plant on the Kaiparowits plateau seemed to be an obvious alternative. Indian land was not involved, deep-mined coal would be available, and there would be local support for the project. Although the estimated construction costs for an oil plant were lower than those for a more capital-intensive coal plant, the cost of fuel for a coal-fired plant was less than oil, making the economics of the two generation alternatives comparable.

The first step that the Kaiparowits project faced was ensuring an adequate water supply. Water would have to come from the state of Utah's Colorado River allocation. The project's application for 102,000 acre-feet/year until 2011 (with a decreasing allocation thereafter) was filed in January 1964 and was granted 21

---

* The utility officials' concern over further development on Indian land was justified. In 1971, the Navajo tribe filed suit against the Department of the Interior for violating their fiduciary duty to the Indians, asking that the remedy be a revocation of the contract. Also, the Hopi Tribe filed suit against the Department of the Interior and Peabody Coal Company, attempting to revoke the lease at the Black Mesa Coal Mine, which supplied 100 per cent of the Mohave Plant coal.

months later, contingent on the commencement of plant construction by 1970.

Initial reactions to the project were very favorable. In 1965, William Gould, an Executive Vice President of Southern California Edison, met with Utah Governor Calvin Rampton at the proposed site. The governor was extremely enthusiastic, seeing the project as a way of aiding southern Utah by providing both jobs and tax revenues (the area had been suffering from serious economic blight since the Depression).

However, obstacles soon appeared to dispel any illusions that the completion of the project would be simple. Although water rights had been acquired from the State of Utah, the water was to be diverted from Lake Powell, a federal area under the control of the Bureau of Reclamation, Department of the Interior. In response to the participants' application for diversion rights, Secretary of the Interior Stewart Udall said diversion rights would not be allowed unless the federal government was allowed to retain control of the flow of water, thereby securing adequate protection for downstream water supply. The controversy between the Department of the Interior and the state of Utah over the control of water rights lasted for four years. Meanwhile, the utilities delayed Kaiparowits in their resource schedules and waited. It was not until October 1969 that a new Secretary of the Interior, Walter Hickel, approved the Kaiparowits application for diversion of water from Lake Powell. At this point, it was clear that a 1970 construction commencement was no longer feasible, and for this reason the participants approached the Utah State Engineer, requesting an extension of their water rights to a 1975 construction date. This request was soon granted.

One additional result of delay was an increase in the estimated cost of the project from the original estimate of $500 million for a 5000-megawatt plant to $750 million for a scaled-down 3000-megawatt plant to consist of four 750-megawatt units. The cost increase was attributed primarily to increases in the price of equipment and labor. In 1969, preliminary project engineering recommenced and the Bechtel Corporation was authorized to proceed with development drilling in order to determine more accurately the coal reserve that would provide the project's 12-million ton/year requirement.

In the early 1970s, the environmental movement grew rapidly, armed with an important procedural tool: the National Environmental Policy Act of 1969 (NEPA).* The increasing number of proposed coal-fired plants in the Southwest drew environmentalist attention to the area. In 1970, the Sierra Club filed a suit to halt any further plant construction in the Southwest until DOI could complete a regional Environmental Impact Statement discussing the cumulative impacts of such construction. Although the suit was later dismissed, DOI took it as an admonition that they would have to pay close attention to NEPA documentation standards in the future.

In response to adverse environmentalist reactions, the participants changed the proposed Kaiparowits site from Warm Creek to a site further from the shore of Lake Powell, Nipple Bench. This site change was approved by the Utah State Legislature in early 1971. Growing environmental concern also led to U.S. Senate–conducted public hearings on generation projects in the Southwest. Secretary of the Interior Rogers Morton responded to these hearings and the surge of Southwest utility development plans by announcing a one-year delay on energy development decisions while a DOI task force considered the long-run environmental effects of Southwestern electric power development and sought a method for achieving development with minimum

* NEPA requires that all major federal actions be accompanied by documentation of the environmental consequences of the action and mandates that federal agencies incorporate environmental inputs and public participation fully into their planning process.

adverse environmental impact. The work of the task force focused particularly on Kaiparowits, the only project immediately pending. The product of the task force, the Southwest Energy Study, concluded that some development, with attendant environmental disruption, was inevitable, but that determinations on particular projects should be made by the Secretary of the Interior.

While the Southwest Energy Study was being completed, the participants were compiling an environmental impact statement for July 1973 submission in order to satisfy NEPA documentation requirements. Nineteen separate sites were examined. At this time, the utilities anticipated quick DOI approval and believed that construction would commence shortly. On June 1, 1973, the existing participants and a new Arizona member, Salt River Project, entered into a Power Participation Agreement, which defined their respective entitlement shares of Kaiparowits power.*

DOI approval was not forthcoming. On June 13, 1973, prior to the participants' submission of their environmental impact report, Secretary Morton endorsed the Southwest Energy Study and denied all pending right-of-way applications for the Kaiparowits project. Citing the results of the Southwest Energy Study as authority, Morton stated that Kaiparowits would impose added severe impacts upon a major recreation area. This decision followed the DOI go-ahead on the trans-Alaskan pipeline, which had characterized DOI as "antienvironmentalist." Many have suggested that the Kaiparowits denial was a concession to environmentalists, made to enhance DOI's tarnished reputation in environmentalist eyes.

---

* Percentages of Kaiparowits megawatts: Arizona Public Service—18 per cent, Salt River Project—10 per cent, San Diego Gas and Electric—23.4 per cent, SCE—40 per cent, unsubscribed—8.6 per cent.

The participants were taken aback. Their response was to go to the White House to try to effect a reversal. Representatives of the participants met with John Love (energy czar) and Charles Di Bona (special consultant to the President on energy), to discuss the merits of the project. On June 19, Governor Rampton, Senator Moss, and Congressmen Owens and McKay (all of Utah) met with Morton, arguing that the project had tremendous economic value for the state of Utah. Two days later, the participants met with officials of DOI, EPA and the National Oceanic Atmospheric Administration (NOAA) prior to the official submission of their environmental report on June 26.

Under pressure from the White House, Morton retreated from his position of absolute rejection and stated that he would review the environmental report and that the DOI would work with the participants in finding a more acceptable site. Morton's negative decision had coincided with the OPEC oil embargo, which increased the price of low-sulfur oil to three to four times its pre-1970 price. This occurrence allowed Morton to justify his reconsideration on the grounds of the Nixon Administration's Project Independence goals.

The utilities proceeded to examine alternative sites in the area. Four site choices were proposed; Morton chose Four Mile Bench. This site was 15 miles north of Nipple Bench and 1000 feet higher, thus presenting less of a visual intrusion problem when viewed from Lake Powell. In mid-December 1973 DOI accepted renewed applications for Kaiparowits at Four Mile Bench (with Nipple Bench as the alternative site) and the companies again proceeded with preliminary engineering work. At this time, the cost estimate had risen to $1.7 billion for the 3000 megawatt project, but spiralling oil prices sustained the economic viability of the project.

In late 1973 the court reinterpreted NEPA to require government-prepared but privately financed environmental impact statements (EIS) for private projects in order to preclude self-serving documentation. The Secretary of the Interior designated the Bureau of Land Management (BLM) as the lead agency responsible for completion of the Kaiparowits EIS. In mid-March 1974 the initial organization meeting was held and the EIS process formally began.

In early 1974 the participants began a public relations campaign in southern Utah in order to encourage support for the project. Public meetings were held in which the participants explained the proposed pollution control methods and stressed the absence of strip mining and the excellent community planning already underway. Kaiparowits was billed as the "biggest and cleanest" coal project ever to be built.

The size and pollution abatement equipment had their price. In January 1975, detailed cost estimates for the project were computed pursuant to a clause in the Participation Agreement allowing any participant to "bail out" before a certain date. Although it was assumed that costs were generally rising, no one was prepared for the results of the detailed analysis. Earlier estimates had been little more than back-of-the-envelope calculations based on Four Corners and Mohave costs.* A more detailed projection showed a cost of $2.7 billion for the generating plant only, with coal costs rising from 19 cents/million BTU in 1971 to 84 cents in late 1974 and $1.22/million BTU at this presentation. (These figures are estimated fuel costs for the year 1982, which was the projected start-up date for the power plant.) With this cost in mind in mid-1975, after reviewing their generation resource options and decreasing demand growth projections, Salt River Project withdrew from Kaiparowits, increasing the uncommitted portion of the project from 8.6 per cent to 18.6 per cent. Not withstanding cost increases and Salt River's withdrawal, the remaining participants signed contracts for turbines and steam generators in

* See Chapter 6 for a discussion of cost estimation.

July 1975. Although the contracts contained standard "escape clauses," they marked a significant commitment to continuation of the project.

In July 1975, the draft EIS was presented to the reviewing agency, the Council on Environmental Quality (CEQ), in 2500 pages and six volumes. One of the most important findings of the draft EIS was that the Kaiparowits power plant would have a "significant effect" on visibility in the area. It stated: "Potentially the most severe visual impact to the area, and possibly to the entire region, could be pollution created by smoke and other airborne particles emitted from the plant." Visibility could be reduced at certain times between 10 and 20 miles in southern Utah and northern Arizona. These comments provided a focus for environmental opposition as did the demand growth predictions contained in the EIS, which showed a 6.8 per cent cumulative load growth figure for the participants' market areas. The FPC had calculated only a 0.5 per cent growth in the market area for the previous year. For the environmentalists, this calculation strongly suggested that the utilities were overshooting the mark in their demand projections.

In September public hearings on the EIS were held in Utah, Arizona, Nevada and California. Opponents of the project questioned the need for Kaiparowits and underscored the harmful effects noted in the EIS. In addition to visibility impact, they stressed increased salinity in the Lower Colorado River, increased mercury content in Lake Powell, and "people pollution" of the area. At the request of the environmental groups who desired a longer period of time to review and comment on the EIS, BLM extended the comment period 45 days, resulting in a projected completion date for the final EIS of March 1976.

Environmentalists were mobilized by the results of their review of the EIS. Between October and November of 1975, nine legal motions, petitions, and official protests were filed in state and federal courts concerning Kaiparowits. These included a procedural suit filed by the Sierra Club with the California Public Utilities Commission and a motion filed by the Escalante Wilderness group with the Utah State Engineer to require formal hearing procedures regarding the Kaiparowits water allocation. This environmental group argued that the power plant was not a "beneficial use" of water, as required by statute, and wanted to remove the Utah State Engineer's discretionary authority.

The National Park Service (NPS) was also motivated to act. It announced plans to conduct a six-month study to provide the basis for a request that EPA redesignate national parks surrounding Kaiparowits as Class I air quality areas.* (This designation would preclude all but minimal degradation of air quality in the area.) The Secretary of the Interior, Thomas Kleppe, allowed NPS to complete only a three-month study of the air quality impacts of Kaiparowits on Bryce Canyon, Capitol Reef and Glen Canyon Recreation Area with the specific purpose of aiding him in his decision.

By December 1975 the companies faced both increasing opposition to the project and an imminent commitment of financial resources. The total project costs were estimated at $3.5 billion (with $600 million for environmental controls) and completion was becoming more and more uncertain. On December 31, SCE's Gould announced a one-year moratorium on financial commitments for the project, pending the DOI decision and other regulatory determinations, stating that "delays in regulatory approvals, stemming from objections by environmentalist groups and lengthy approval processes have forced deferral for one year."

Events in early 1976 catapulted Kaiparowits to the national scene. In addition to national media coverage in all the major newspapers, Robert Redford appeared on a March *60 Minutes* television program and presented an emotional plea against the

---

* Appendix A gives a brief history of the Clean Air Act.

project. Despite widespread vocal opposition to Kaiparowits, however, Utah public opinion remained positive. A group of southern Utah citizens, in a bus partially subsidized by the participants, drove to Washington to meet with congressmen and federal officials and express their support.

In March, the NPS study, "Analysis of Kaiparowits Powerplant Impacts on National Recreation Resources," was released. Its well-publicized conclusion was that pollution from the plant could be visible for 100 miles. After the fact, the NPS study methodology has been heavily criticized, but its immediate impact was significant. Also in March, BLM submitted the final EIS to CEQ, a document that cost the participants $1.2 million to complete. After initial review, the EPA announced that it would advise the Secretary against an affirmative decision on the project, questioning the need for and effects of the plant.

Two factors external to the Secretarial decision also had the potential to affect the history of Kaiparowits. The California Nuclear Initiative to be determined in June was designed to halt nuclear development in the state, increasing the California participants' stake in Kaiparowits. At the same time, Congress was considering a set of amendments to the 1970 Clean Air Act that would automatically designate national parks and surrounding areas as Class I air quality areas, resulting in such tight restrictions on Kaiparowits that its construction and operation would be put in doubt.

Secretary of the Interior Kleppe was to visit the proposed plant site on April 21, 1976, prior to making his decision on the project. Before that time, he met individually with all concerned groups so that they could express their diverse views on the project. Various environmental groups met with Kleppe and stressed both their objections to potential air quality effects and to the demand growth estimates. A political delegation from Utah, including Governor Rampton, stressed the benefits to the state as did the

Utah citizens' delegation who discussed for three hours the problems of unemployment and forced outward-migration of youth.

By April 13, however, the participants had come to the decision to cancel Kaiparowits. Gould and SCE's Washington counsel, Alan Nedry, attempted to see Kleppe again to inform him of this decision. Kleppe's theory of "equal access" prior to decision-making led to his refusal of a meeting, and they instead informed others in DOI who relayed the message. The decision was made just prior to the time that Governor Rampton was to make a TV appearance supporting the project. Cancellation was formally announced by SCE and SDG&E the following day and by APS on the 15th.

At the time of cancellation, the participants had expended 13 years and $20 million on Kaiparowits in a frustrating endeavor to get the project to construction. Gould stated that "rising costs, regulatory delays, and lawsuits" played an important role in the cancellation but that financing was also a crucial concern—whether the participants could justify diverting today's wealth to cover tomorrow's needs, not knowing what the ultimate cost might be in the face of the delays that could be foreseen and those that could not be.

Two days after the cancellation, southern Utah citizens held a public protest meeting to lynch in effigy those they felt were responsible for the demise of the project.

Although the probabilities of a coal-burning project on Kaiparowits plateau in the future look low, the valuable local coal reserves are still there untouched. In May 1976, the participants completed a brochure on the potential of a coal gasification project using their Kaiparowits leases, hoping to attract interest for such a project.

3
The Environmental Dispute

The release of the Kaiparowits environmental impact statement (EIS) on March 3, 1976, engendered two diametrically opposed responses. Project participants insisted that the EIS showed that Kaiparowits' effect on southern Utah would be minimal and acceptable—"The integrity of the environment will be maintained," they assured. The environmentalists bitterly disagreed. Their response, in reference to the same EIS projections, was perhaps best exemplified in an article in the March 1976 *Audubon Magazine,* the author calling Kaiparowits "the ultimate obscenity"—depicting it as a serious threat to some of the nation's greatest parks and canyonlands. Environmentalists also contended that the need for Kaiparowits had never been independently verified and that electricity conservation potential in the Southern California and Arizona market areas had never been seriously considered as an alternative to Kaiparowits' construction.

This chapter will review the major findings* of the Kaiparowits EIS and will highlight the divergent reactions to it. The chapter

* Only those findings that sustained the most debate will be presented.

will show that the plant's environmental impacts, and the need for the plant in the first place, were subject to great uncertainties —uncertainties that could not necessarily be resolved through increased methodological sophistication. The methodological limitations of environmental and energy demand forecasting, together with the differences in the interpretation of forecasted results, may significantly limit the formulation of effective energy and environmental policies.*

FINDINGS OF AND RESPONSES TO THE EIS

The Kaiparowits EIS provided great detail regarding a wide range of possible environmental consequences of Kaiparowits' operation. The report included discussions of possible effects upon:

- climate
- air quality
- geology
- soils
- water resources
- vegetation
- wildlife
- paleontology
- archeology
- history
- recreation
- land use
- socioeconomics

* A technical evaluation of the formal models used to project Kaiparowits' environmental impacts is beyond the competence of the authors of this study. We will present some of the analytic comments that were made regarding the modeling.

The controversy over Kaiparowits focused on the effect of the plant's emissions on visibility in the area; the extent to which the facility itself would represent a physical intrusion on the scenic landscape or have negative impacts on the quality of the area's limited water supply were the other two issues that sustained the most debate. These latter two issues, however, clearly attracted less concern than did the potential visibility reductions on and around the Kaiparowits plateau.

Visibility

Combustion of coal for the production of electricity produces a number of pollutants, among those being: particulate matter (including fly ash, sulfates, and nitrates), sulfur dioxide, nitrogen oxide, methane, mercury, and small quantities of elements such as arsenic, barium, boron, fluorine, lead, thorium, selenium, and vanadium. Increased levels of particulate matter in the atmosphere and the conversion of sulfur and nitrogen oxides into aerosols reduce visibility in the area affected.

To reduce such emissions, utilities have incorporated expensive and complicated control devices into their plant design. Kaiparowits participants submitted that their plant would have the most sophisticated pollution control equipment available. Electrostatic precipitators, which use electric fields to capture particles in the flue gas stream, were designed to remove 99.5 per cent of all particulate matter. A wet lime scrubbing system, which forces flue gas into a chamber where sulfur dioxide interacts with hydrated lime and is removed, was to attain a removal rate of 90 per cent.

EIS writers were aware that operating coal plants were not achieving this level of pollution control. In fact, only 2 of 15 coal-fired plants in the East with electrostatic precipitators were meeting their pollution control design specifications. The Navajo coal plant in Arizona was achieving 96 per cent particulate removal.* Four Corners was achieving 97 to 98 per cent removal. The Huntington coal plant (430 megawatts), 150 miles north of Kaiparowits, was at 99.5 per cent efficiency. The EIS held that the level of pollution control foreseen for Kaiparowits was probably feasible by the time that Kaiparowits would be under construction [2].

Achieving this level of pollution control, burning average quality coal, and operating 75 per cent of the time in a given year, Kaiparowits would annually release 2820 tons of particulates, 7555 tons of sulfur dioxide, and 68,000 tons of nitrogen oxides, according to EIS projections [3].** These emission levels would be equal to or below the applicable Class II area standards established by state and federal governments in compliance with the Clean Air Act,† even under the "worst case" meteorological conditions. If some of the nearby federal park and recreation lands were reclassified as Class I areas, where practically any negative change in air quality is disallowed, emissions could possibly exceed applicable standards (preliminary model predictions did exceed such standards).‡ No human health effects were anticipated at the projected emission levels.

The state-of-the-art of emission modeling has advanced greatly in the recent past. However, even though the models used are now quite sophisticated, the estimation process remains quite inexact. Environmental scientists do have some empirical under-

---

* More specifically, it was achieving 99.1 per cent removal, with this pollution control equipment available 97 per cent of the time.

** See Appendix B for further emission details.

† In Class II areas, air standard values allow a deterioration of air quality "normally accompanying moderate, well-controlled growth." Appendix C details the standards that Kaiparowits would have had to meet. Appendix A reviews the history of the Clean Air Act.

‡ SCE officials have stated that some of their studies suggested that Kaiparowits could meet Class I standards all but 80 hours of the year.

standing of the degrees of emission associated with the combustion of different quality coal. The effects of these emissions are dependent on the pattern and the degree of dispersion of the so-called emission "plume" as it leaves the stack. This dispersion is determined by complex meteorological conditions (wind currents, for example), and by the topology of the area. These conditions are very difficult to model. In order to simulate the dispersion of pollutants under stable and unstable atmospheric conditions, EIS consultants conducted smoke releases at various sites of the Kaiparowits plateau.

Using such simulations, EIS consultants were able to estimate the potential visibility impacts of the project. The central conclusions were developed by the Bechtel Corporation in a study sponsored by the participating utilities.* The EIS summary of potential impacts is worth quoting *en extenso* since the ensuing debate was in reference to these conclusions:

> First, the distance one can see could be substantially reduced. . . . Visibility in the scenic region of southern Utah and northern Arizona could be reduced by 10 to 20 miles, which would be an adverse effect on the viewer's perception of the open characteristics of the landscape and negate the more distant topographic features. A study (Bechtel Power Corporation—1974) sponsored by the participants indicates that reduction of visibility of this magnitude (i.e. 10 to 20 miles) would occur infrequently.
>
> Second, coloration of pollutants could alter the natural coloration of blue sky and particularly affect the contrast of scenic topographic features against the skyline. The de-

* Although the EIS includes almost 100 pages of technical reference material on air quality studies, the connections between the Bechtel study and the plume dispersion studies in developing visibility impact conclusions were not clearly stated.

> gree of impact would depend on the viewer's position in relation to the plume.
>
> Should any of the pollution control equipment malfunction, there would be a possibility of severe visual impacts. There would be a good possibility that dense visual air pollution would drift into outstanding scenic areas such as: Bryce Canyon National Park, Glen Canyon National Recreation Area, Dixie National Forest, Kaibab National Forest, Grand Canyon National Park, Zion National Park, Capitol Reef National Monument, Monument Valley and many other highly scenic areas. . . . The probable frequency and magnitude of occurrence of a visible plume or haze are not well defined. . . . If the pollution control equipment operates at design efficiency, it is highly unlikely there would be a visible plume. However, technology is not as far advanced for controlling the yellow discoloration (haze) associated with nitrogen oxide emission. Hence the probability for haze is greater. Again, the probable magnitude and frequency are not well defined. Observations at other generating plants having similar pollution control equipment as that proposed at Kaiparowits are mixed. A yellow discoloration has been periodically observed for the first Navajo unit and has deepened with the start up of the second unit. The experience at the San Juan and Huntington plants seems to be better. The potential does exist for periodic high visual impacts due to visible plume or haze conditions resulting from stack emissions [4].

The EIS concluded that "emission of free particulate and conversion of sulfur and nitrogen oxides to aerosols would cause a visibility reduction that could not be mitigated" [5]. "The loss of clean air during the project life would be irretrievable, particularly to Glen Canyon National Recreation Area and Bryce Canyon and Capitol Reef National Parks" [6]. "Haze and sky discoloration from plant emissions would adversely affect the scenic resources of south-central Utah and north-central Arizona" [7]. Because of the remoteness of the area, however, the numbers of people who would be affected by the periodic visibility reductions would not be large. The EIS also noted that the cumulative effects of the Navajo and Kaiparowits plants could cause "rapid degradation of existing air quality" and that national ambient air quality standards could be exceeded in areas distant from the plants [8]. Other studies cited in the EIS argued that the probabilities of cumulative impacts of Navajo, Kaiparowits, and the five other coal plants within 200 miles would be small "considering the separation distance, relationships of plant sites to prevailing winds, large atmospheric dilution potentials, interposing terrain features and types of emission controls proposed" [9]. The data necessary for a reliable projection, however, was not available.

In their "Participants' Position Statement," which detailed arguments for approval of the project, the utilities stressed that Kaiparowits emissions would be less than allowable levels as determined by Utah and the federal government. The document summarized the EIS' visibility findings as a basis for assuming that the "integrity of the environment would be maintained" [10]. Utility officials have also suggested that the emission dispersion models used for projecting visibility impacts tended to underestimate the degree to which wind currents break up the emission plume and may therefore have overestimated visibility impacts. Utility officials have argued that the visibility impact projections should be seen only as worst-possible-case scenarios and stressed that such impacts would rarely occur.

Environmentalists, however, saw something completely different in the EIS figures and tables regarding air quality effects. The aforementioned *Audubon Magazine* article claimed, in no uncertain terms, that Kaiparowits could create, in conjunction with the other plants already built in the region, "a permanent haze over the entire Lake Powell-Grand Canyon region. . . . converting this heartland of the scenic Southwest into the most polluted nonurban area in the United States" [11].

Criticisms of the EIS findings tended to focus on four topics. First, environmentalist critics called attention to the past record of coal plants in the Southwest, particularly the Navajo plant. The Environmental Defense Fund (EDF) noted the comments in the draft EIS indicating that average visibility in the Kaiparowits region prior to the operation of the Navajo plant was 90 miles; at the time of preparation of the EIS, visibility averaged 70 miles [12]. Friends of the Earth (FOE), in a letter to Senator Jackson (Chairman of the Interior Committee), asserted that the pollution from Navajo was occasionally bad enough to obscure the north rim of the Grand Canyon from an observer looking from the south rim [13]. The Audubon article quoted a Utahn referring to a cumulative buildup of haze attributed to Four Corners and Navajo, obscuring the canyonland panorama [14].

Second, FOE also questioned whether Kaiparowits could reach and maintain a design efficiency of 99.5 per cent particulate removal. A reduction of efficiency by 2.5 per cent (to the operation efficiency of Navajo) would increase the level of particulate emissions sixfold [15].

Third, FOE suggested in their comments reviewing the EIS that the visibility effects of Kaiparowits projected in the EIS would be serious enough, but that the potential visibility impact might also have been underestimated. FOE held that the plume dispersion simulations conducted for Navajo, similar to those completed for the Kaiparowits EIS, yielded emission estimates that were in error by a factor of fifteen (no further explanation or reference was given) [16]. FOE noted a Denver Environmental Protection Agency office calculations suggesting that 24-hour sulfur dioxide concentrations at the south end of the Kaiparowits plateau would exceed applicable standards, a conclusion that conflicted with that of the EIS [17]. EPA itself noted what it took to be other modeling problems in the EIS. For example, EPA stated that the particulate removal efficiency of 99.5

per cent was by mass, not number, of fine particulates, and that these fine particles are a primary cause of reduced visibility [18]. EPA also believed that the haze produced by emission of nitrogen oxides could be greater than that produced by Navajo, due to Kaiparowits' larger size [19].*

Fourth, environmentalists stressed that the cumulative effects of a number of existing and proposed coal plants on air quality in the Southwest could be devastating and that the cumulative impacts necessitated more study than that undertaken for the EIS. Dr. O. L. Anderson, in testimony before the California State Energy Commission in November 1975, argued that the combined emissions of Navajo and Kaiparowits could equal one-half of the ash emissions, one-half of the NO_x emissions, and one-third of the SO_x emissions in the entire Los Angeles basin [20]. Environmentalists argued that a comprehensive regional EIS assessing the cumulative impact of all existing and proposed plants within the Colorado plateau would be necessary before making the final decision on the plant.

The environmentalists found what they took to be significant support when the National Park Service (NPS) released its 47-page "Analysis of Kaiparowits Powerplant Impacts on National Recreation Resources" in March 1976. The report indicated a good deal of agreement with the environmentalist objectives. The study was an analysis of data that was already available, much of it in the Kaiparowits EIS. NPS' air quality and visibility conclusions were based on atmospheric dispersion modelling prepared by the Environmental Protection Agency (EPA) for NPS.

The study stated that Four Corners and Navajo produce yellow-brown or reddish haze that seriously interfaces with visibility over

---

* EPA commented on the Kaiparowits EIS in accordance with its environmental review responsibilities under section 309 of the Clean Air Act.

100 miles from the plants [21].* Kaiparowits operation, the study argued, would lead to a "one-third to full reduction" of visibility from places in Bryce Canyon National Park at least 15 per cent of the time; this degree of reduction would occur in Glen Canyon National Recreation Area at least 14 per cent of the time [23]. Moreover, there would be occasions when the plumes from Kaiparowits and Navajo would affect the same areas, leading to "synergistic impacts" (left undefined) [24]. In general, the NPS report was a presentation of conclusions, without much detail regarding the baseline data, theoretical analysis, and computer simulations used. Because of this weakness, other government agencies found the report difficult to endorse. After the publication of the report, however, the environmentalists redoubled their efforts against Kaiparowits.

Physical Intrusion

The NPS report also placed much more stress than did the EIS on the extent to which the project would intrude on the Kaiparowits plateau landscape. The EIS had stated:

> The greatest potential visual impact would likely occur at Bryce Canyon National Park. The plant would be visible from most of the park overlooks. Visual impacts would range from low to very high. If only the plant structures are visible, the impact would be low. . . . since the plant would be 32 air miles from the nearest overlook and would be backgrounded by the Smoky and Fifty-Mile mountains. However, if a stack plume is visible or a haze condition develops

---

* The study included four photographs showing seemingly significant visibility effects of the Navajo plume. Opacity attributed to the Navajo plume was quite visible, and clearly obscured the Glen Canyon and Bryce Canyon scenes [22].

> from the stack emission which would obscure landscape features and detract from the natural beauty of the area, the visual impact could be very high. Other conditions which may contribute to the high impact are the lights at the plant (i.e. strobe lights on the stack during daylight hours and vehicle and plant lights at night) and reflections from vehicles and plant windows or other reflective materials [25].

NPS went farther to argue that within 60 miles of the plant, "the vertical rise of the plume would be seen from 11 per cent of the area, the stack would be seen from 8 per cent of the area, and the power block would be visible from 5 percent of the area" [26]. The plant, states the study, would represent "a physical intrusion directly visible from 11 of the 12 key observation points within Bryce Canyon National Park" [27], as well as from other observation points in the region.

The NPS report concluded that the operation of Kaiparowits would "substantially reduce visitor appreciation of scenic resources within 60 miles of the plant" [28] and cautioned that "maintenance of clean air, allowing long-range, unobstructed viewing, is essential to provide a satisfactory visitor experience for future generations" [29].

Water Quality Effects

Although visibility and physical intrusion were the prominent issues in the eyes of the environmentalists, Kaiparowits critics were also concerned about a number of other potential project impacts detailed in the EIS, among those being water quality effects. Water rights and water quality issues have long been touchy subjects in the arid Southwest. As mentioned in Chapter 2, for example, the Kaiparowits project was stalled during the 1960s because of an ongoing debate between Utah and the federal government over the control of water resources.

Any water withdrawal from the Colorado River increases the salinity of water for downstream users. The EIS stated that withdrawal of the 50,000 acre-feet of water required by the plant for cooling water would increase salinity by an estimated 2.1 milligrams per liter (mg/1). For each mg/1 increase in salinity, it was estimated that an annual damage of $230,000 could occur in terms of agricultural, municipal, and industrial use [30]. Withdrawal of the water would also make the water unavailable for other uses and would reduce Utah's remaining share of Colorado River water by about ten per cent.

Also, mercury emitted by the plant would be deposited on the ground by rainfall and move into the Lake Powell system. The large game fish of Lake Powell already contained mercury concentrations exceeding the maximum safe levels for human consumption set by the federal Food and Drug Administration (FDA). Even small increases in the mercury levels of the water could cause much greater increases of mercury levels in these fish because mercury levels are amplified in the food chain. The EIS cautioned that the rate at which mercury returns to the surface had not been well documented, but estimated that the amount would be between 1 and 27 per cent of the estimated yearly mercury deposition into Lake Powell. With the operation of the Navajo plant, the total mercury input would be from 9 to 35 per cent of the annual natural deposition [31]. The EIS concluded that "[A] significant increase in mercury levels of game fish would cause some degree of loss to the Lake Powell sports fishery. The degree of loss could range from a slight lowering of quality to complete loss of fishery to public use depending on the degree of contamination" [32]. In their position statement, the utilities cited studies suggesting that mercury emissions would not even reach the atmospheric saturation level necessary before there would be absorption into the water system [33].

To a large extent, project proponents and opponents were talking past one another. Regarding the major issue, air quality, the

EIS had stated that there would be an unavoidable negative impact, although stressing that the magnitude of impact could not be determined with great precision and that any such impact would not affect great numbers of people because of Kaiparowits' remote location. The utilities essentially asked that the reviewer accept their studies as an adequate basis for concluding that any such projected impacts would turn out to be unimportant, and argued that the EIS probably overestimated impacts and definitely gave a misleading impression of the probabilities of impact by emphasizing "worst-case" scenarios. Environmentalists, on the other hand, vehemently argued that the EIS results showed that Kaiparowits promised continuing deterioration of treasured Southwest canyonlands and that the EIS impact estimation might even be conservative.* The differences in interpretation of the detailed document are troubling, indicating not only the primitive state of environmental impact forecasting, but also the lack of anything approaching a resolution of the Kaiparowits debate.

Similar uncertainty was reflected in the controversy over the need for Kaiparowits, to which we now turn.

THE NEED FOR KAIPAROWITS

Energy demand projections, the causal relationships between energy expansion and energy use, and energy conservation po-

---

* It is interesting to note that EPA, in its EIS review capacity, believed that Kaiparowits was "environmentally unsatisfactory as proposed," and that approval of the project might be inconsistent with the legislative mandate of the Park Service (protection of the national parks for use by future generations) [34]. However, they stated that their review was tentative because it was constrained by the EIS' "unsatisfactory" analysis of air quality effects and mercury emissions.

tential have become subjects of increasing controversy over the last decade. The methodological problems with demand projections are obvious: any such projection makes a number of assumptions regarding economic growth, demographic growth, energy prices, and conservation measures; these assumptions are necessarily based upon past trends that may change radically in the future. In the case of Kaiparowits, the participating utilities held that Kaiparowits was necessary to provide an adequate power reserve margin* in the 1980-1985 period; environmentalists argued that the need for Kaiparowits was never conclusively shown, and that conservation was not taken seriously enough as an alternative to Kaiparowits.

No one had anticipated the significant changes in U.S. electricity consumption patterns that were the result of the 1973 OPEC oil embargo and the economic recession of 1974-1975. After 15 years in which national electricity demand grew at an average annual rate of 7.4 per cent, growth dropped precipitously in 1974. SCE's customers' average annual consumption of electricity grew at a compound rate of 7.9 per cent from 1964 through 1973. In 1974, consumption declined by 5.6 per cent and by the end of 1975 was still below the 1973 level. Changes in demand levels shook the electric utility industry in two ways. Such changes penalized earnings by forcing many companies to produce at below normal levels, thus reducing revenue while fixed charges

* A utility's reserve margin measures the difference between its total generation resources and its expected peak load demand. Since equipment cannot function without some chance of failure, an electric utility is never 100 per cent reliable. There is always a fraction of a utility's generation capacity that cannot be used because of scheduled maintenance and unexpected technical problems. The frequency of service interruptions due to generation failure in a system depends on the quantity of reserve margins available. Higher generation reliability requires higher reserve margins.

could not be reduced. Beyond that, these changes threw both utility executives and governmental regulators into a state of uncertainty regarding long-term demand for electricity.

In their EIS submission, the utilities projected that electricity demand would return to an annual growth rate of 5.5 per cent in the participants' market areas with the resumption of normal economic growth in 1976 (three to four per cent annual increase in GNP) [35]. Decreasing availability and increasing cost of natural gas would also increase the demand for electricity. Kaiparowits' 3000 megawatts would have represented 26 per cent of the generating new capacity added in the 1975-1985 period by the three participating utilities. The utilities wanted to increase base load capacity* for economic and reliability reasons. The utilities believed that other technologies, such as solar or wind power, would not be feasible as base load sources in the near future, and that investment in conservation systems was not a realistic or justifiable alternative to Kaiparowits. In utilizing low-sulfur coal, Kaiparowits would lessen dependence on foreign oil, displacing about 80,000 barrels a day [36]. Because of the decreasing availability and increasing cost of natural gas, and the scheduling uncertainty connected with nuclear power plants, the utilities argued that use of coal would be vital for a good resource mix.

In its study regarding the need for Kaiparowits prepared for the EIS, the Federal Energy Administration (FEA) supported the use of Kaiparowits coal. "FEA would support the use of Kaiparowits coal for electric power generation even if growth in demand were less than forecast as a result of increasing costs, government regulatory policies, and consumer conservation practices" on the grounds that Kaiparowits would displace oil consumption

---

* Base load plants, such as facilities, coal, nuclear, and hydroelectric, operate at or near their full capacity during all of the hours that they are available and offer low energy costs per kilowatt hour of generation.

[37]. As alternatives to a plant at Four Mile Bench, FEA mentioned transport of Kaiparowits plateau coal to a plant site in Nevada or eastern California, where environmental impacts might be less significant. The problem with such an alternative site, the study continued, is that it might be less suitable in regard to water availability, ability to meet air quality standards, or the economics of coal transportation. Coal gasification or liquefaction, "unproven and less energy efficient, but probably viable technologies," could merit comparison to Kaiparowits on a cost-benefit basis.

EPA focused its attention on the Kaiparowits electricity demand analysis, which it held to be inadequate. The EIS, EPA stated, did not consider the effects of California conservation measures that were being implemented, such as mandatory insulation standards, partial bans on electric heating in new commercial buildings after 1977 (unless it can be justified on a life cycle cost basis), appliance standards, and commercial lighting standards. The EIS also did not consider the possible demand-reducing effects of peak-load pricing, according to EPA.* EPA estimated that peak-load pricing, in conjunction with other load management techniques, could reduce by 1985 the demand for capacity by 13–23 per cent for Southern California Edison, by 9–20 per cent for San Diego Gas and Electric, and by 14 per cent for Arizona Public Service. EPA also held that the alternatives to delaying Kaiparowits by accepting lower levels of system reliability or achieving higher reliability levels were possible. EPA concluded that an independent analysis of the demand for electricity in the Kaiparowits market area was necessary before a proper evaluation of the need for Kaiparowits could be completed [38].

---

* That is, increasing the price of electricity during times of peak demand so that electricity usage becomes more uniformly distributed throughout the day (overall peak system demand is thus reduced, and therefore capacity requirements are lessened).

Friends of the Earth (FOE) led the environmental opposition by criticizing Southern California Edison's forecasting methods. FOE held that there was a shift away from energy intensiveness in the industrial sector of SCE's marketing area, one that FOE believed was inadequately reflected in SCE's forecasting. Nor did the SCE model take into account possible demand-reducing effects of electricity price increases, voluntary conservation, or the spread of solar heating and cooling, FOE asserted. FOE cited California Energy Commissioner Dr. Ronald Doctor's claim that SCE's projections in fact assumed that electricity prices would decrease relative to other prices [39].

In this regard FOE cited FEA's statement that "[f]inancial incentives and disincentives combined with regulations concerning electrical power use could limit electric power demand to the point where new generation capacity would not be required." Ronald Doctor directed attention to an earlier energy demand study that he had conducted for the Rand Corporation, which had projected effects of rising electricity prices and conservation efforts about or likely to be implemented. For the SCE market area, this study projected a 1984 peak demand that was 2440 megawatts below the SCE figure, enough power to displace twice SCE's share of Kaiparowits.

The California Governor's Office of Planning and Research expressed concern over the relationship of Kaiparowits and the future of conservation efforts in Southern California, a concern that was paramount in the thinking of environmentalists: "if demand forecasts overestimate the actual electricity need and a surplus of electric energy is available, growth will be encouraged to fulfill the demand forecasts and proposed conservation efforts will be curtailed" [40]. Kaiparowits supporters did not feel that it was meaningful to talk of self-fulfilling forecasts. They held that forecasts were based on empirical and well-founded assumptions, and stressed that it is not in the self-interest of utilities to overbuild capacity. The costs of large facilities are high, and financing is no small task. Utilities would also likely argue that demand is not dependent on the size of the reserve margin. Moreover, aside from advertising electricity usage, which would be politically taboo in the mid-1970s, utilities possessed no means of encouraging demand, even if they wanted to. Critics could point to some utility practices such as price-breaks for large industrial users, but this example would not provide adequate basis for the "self-fulfilling forecast" argument, since such prices are set by the public utility commission, not by utilities themselves.

Participants, government officials, and environmentalists were talking on different levels when they spoke of the "need" for Kaiparowits. If Project Independence goals of reducing dependence on foreign oil were paramount, then the use of Kaiparowits coal looked good independent of what demand projection showed, and the plant was "needed," as FEA held. If Kaiparowits was considered to be a serious threat to the environment, then it was imperative to forego Kaiparowits and to search for more environmentally benign generation methods, while initiating strong conservation measures. Thus, positions on the "need for Kaiparowits" were determined by social objectives as much as by forecasts of future demand.

4
Environmentalists and Kaiparowits

Environmentalists were quite successful in mobilizing their resources and pushing their position on Kaiparowits. Environmentalists were particularly effective in gaining press coverage of the Kaiparowits issue and in raising some doubt in the minds of congressmen and Department of the Interior decision-makers about the desirability of the project. Utility officials and project proponents claimed that environmentalists waged a war of distortion against Kaiparowits, deceiving the public while using litigation and the threat of litigation to force delay and to increase the project's costs. It is important to review the approach to the environmentalists to Kaiparowits because opposition to Kaiparowits is one significant example of the effect of the environmental movement on energy policy.

The environmentalist position on the project is relatively simple to state. Environmentalists argued that the true alternative to massive development of polluting, coal-fired power plants in the Southwest was energy conservation in the urban load centers. The Clean Air Act, the Supreme Court has ruled, prohibits significant deterioration of air quality in clean air regions in order to prevent utilities and industry from escaping the effect of stringent urban air-quality standards by fleeing to less populated areas. In addition, the environmentalists held that siting plants in such locations amounted to a subsidy from the people of the Southwest to the users of electrical energy in Southern California, who want to consume electricity without paying its true social costs, including pollution. And, environmentalists averred, so long as utilities have the option to build polluting generating plants in the Southwest, they will lack the incentive to develop new technologies for burning fossil fuels cleanly, or for using such sources as solar, wind, or geothermal energy to meet energy needs. In addition, consumers should be educated as to the need for reducing their use of energy, and where necessary, such consumption should be regulated, and limited by appropriate legislation.

No such coal-fired plants, environmentalists believe, should be built anywhere near national parks and monuments, or any other scenic, historical, or sacred site. Friends of the Earth (FOE) adduced the 25 per cent increase in visitation to Grand Canyon between 1974 and 1975 and the 43 per cent increase for Bryce Canyon during the same period as examples of the increasing value of the Southwest parklands to the American people [41]. In addition, FOE noted the Federal Energy Administration poll in which 94 per cent of those surveyed felt that areas having clean air "should be kept as clean as they are now" [42]. Approval of the project, environmentalists claimed, would represent a break with the Congressional charter of the National Park Service, the 1916 Act directing the Service to "conserve the scenery and the natural and historic objects and the wildlife therein . . . in such a manner . . . as will leave them unimpaired for future generations."

Environmentalists were quick to remind that Secretary Morton had initially rejected the plant because he believed that the project represented an environmental gamble. "The scenic beauty of [the] rugged Southwest landscape, coupled with the clarity of air in the vicinity, are national assets of major importance, worthy of protection for the enjoyment of future generations," the Secretary had stated [43]. In letters to and meetings with Secretary Kleppe, environmentalists argued that the risks of Kaiparowits had become even more apparent since the Morton decision and urged Kleppe to deny the project's permit applications.

Environmentalist opposition to Kaiparowits began in the early 1970s, with Sierra Club testimony during the initial siting hearings. As a result of the Secretary of the Interior's recommendations and the Southwest Energy Study results, the utilities began a search for another site. However, the Sierra Club was not satisfied with the newly proposed sites and sustained opposition to the plant began in late 1973.

John McComb, who was primarily responsible for directing the Sierra Club's fight against the project, states that the campaign against the plant was composed of three elements. First, effective opposition required as much national publicity as possible in order to counteract and override the local support of the project. Second, it was essential to compile information regarding the actual effects of the plant, the relevant decision-making channels, and the perspectives of the decision-makers on the project. Third, policies were developed to foster popular support and to enable successful lobbying with decision-makers in the Department of Interior, the Environmental Protection Agency, the National Park Service (NPS), and individual Senators and Congressmen.

In 1975 it became obvious that the visibility of environmental opposition would have to increase if it was to have any effect. A meeting was held in Denver among the leaders of various environmental groups to discuss strategy, but internal strife prevented the development of any formal structure for cooperation. The first major cooperative attempt to focus national attention on the plant was the Lake Powell press conference held in January 1976. The idea for the conference grew out of a suggestion made during the Sierra Club National Board of Directors meeting. Seven national environmental groups called a press conference on the shores of Lake Powell near Wahweap, Arizona, close to the proposed Kaiparowits site. Local citizens, who vigorously supported the plant, discovered the conference plan the night before it was scheduled to be held, and were infuriated that no local representative had been invited to attend. Taking matters into their own hands, they invaded the conference, interrupting the scheduled speakers and demanding equal time on the platform. This direct confrontation between the two groups greatly added to the excitement of the meeting, and the story was covered by most of the national papers.

Media attention is essential to the environmentalist cause. The national media are generally quite receptive to environmental stories. In the abstract, at least, "the environment" is an apple pie–and–motherhood issue—few people are against environmental protection in principle. Media exposure not only helps environmentalists to raise the saliency of an issue, but also enables them to convey their perspective to decision-makers. Initially, much of the information used by the press on an environmental story is provided directly by a staff member of an environmental organization who may have some expertise on the topic and is also more than willing to prepare information and draft articles that will result in sympathetic coverage. When the issue becomes prominent, the press may begin to respond on its own. Environmentalists believe that the press is quite legitimately trying to fulfill the need for airing environmental viewpoints, arguing that environmental quality issues would otherwise go unconsidered, as they had in the past.

Almost all of the environmentalists opposing Kaiparowits were satisfied with their access to the press. Ron Rudolph, a Friends of the Earth representative, stated that the press would come to him for "the environmental point of view," once he had become thought of as a fairly reputable source of information. Environmentalists have been able to use this access to the press quite effectively. If a Congressman, for example, is reconsidering his proenvironmental position on a popular issue, environmentalists will often notify the press, putting significant pressure on the Congressman not to change his mind. Chris Goddard of the National Clean Air Coalition (NCAC) in Washington stated that there were certain writers from several national newspapers and magazines who were very sympathetic to the environmental viewpoint and who readily accepted position statements and information from the NCAC.

A number of utility and government officials were quite disturbed by what they considered to be environmentalist "domination" of the press. Senator Jake Garn's aides (R-Utah) believed that misinformed Congressmen relied on what the aides saw as inaccurate or one-sided newspaper articles to decide upon important environmental or energy policy questions. The March 21, 1976, airing of the segment of CBS' *60 Minutes* devoted to Kaiparowits was highly criticized as being a one-sided and distortion-ridden treatment of the Kaiparowits issues. Robert Redford, who appeared on the program and objected to the project on environmental grounds, was unhappy himself with the show, saying that it "appeared to be biased." He added: "it would have been better for me and the show in general if it had had more balance." The plant site was not shown on the program; the scenery actually shown (but not identified), according to a utility spokesman, included Shiprock, New Mexico, 150 miles away. The potential impacts of the project were discussed only in vague generalities. The proproject opinions of local residents were not aired (Redford had urged that their views be sought and had supplied phone numbers to the CBS crew). The Department of the Interior (DOI) said government agencies received "thousands of letters" after the airing of the segment; Utah Congressmen said their mail, which previously had been heavily favorable to the plant, afterwards ran against it [44].

Another example often raised to suggest environmentalist-induced media distortion had to do with the NPS study. An early draft of the report had included a claim that noxious odors would emanate from the plant and would be evident to park visitors up to 60–100 miles from Kaiparowits. This claim was given extensive media coverage. Few stories included the criticisms of this point and methodological reservations voiced by other government agencies [45].

The Sierra Club, in its attempt to package EIS findings in a way that would have highest visibility, occasionally lapsed into misrepresentation. One publication stated that Kaiparowits would reduce visibility from the "current 85–100 miles to occasionally as low as 15 miles over Bryce and Grand Canyon national parks and Glen Canyon National Recreation Area;" project approval would lead to "pollutant concentrations beyond those permitted in industrial areas of California, Japan, Czechoslovakia, Canada, and Russia will blanket the parks of the region" [46]. The first statement does not reflect the visibility reduction conclusion made in the EIS. In regard to the second statement, it is possible that there are locations within the mentioned areas where Kaiparowits emissions would exceed standards; but as is, the statement is pure hyperbole. The press will often print such statements as newsworthy in themselves, without checking for their accuracy.

Environmentalists have admitted to stretching the truth occasionally. Ron Rudolph, however, insists that the number of environmentalists who intentionally misled the press was small. Environmentalists believed that the utilities distorted evidence when they continually assured people that "environmental integrity" would be maintained, thus discounting the type of significant impacts that could have actually occurred.

Environmentalists were also successful in lobbying against Kaiparowits at the national level. In Congress, environmentalists are often considered to have greater credibility than industry spokesmen as data sources, although their evaluation of the data may be suspect. Environmentalists tried to convey their objections to the project to Congressmen and Congressional staff and to the Assistant Secretaries in the DOI. Friends of the Earth had allies such as Congressman Timothy Wirth (D-Colorado), who sent a letter to his colleagues in late March 1976, urging them to ask Secretary Kleppe to postpone his decision on Kaiparowits until an independent energy demand assessment and the Clean Air Act deliberations could be completed.

The National Clean Air Coalition (NCAC) in Washington pressed for redesignation of national parks to Class I status in the pending Clean Air Act amendments. The NCAC was originally organized in 1973 to work for an amendment to the Clean Air Act of 1970. Although the NCAC has approximately 45 members, including the Sierra Club and Friends of the Earth, decisions regarding clean air legislation were made by NCAC leadership and passed down to the members. NCAC had excellent Congressional access and good press contacts.

Environmentalist activities were much more limited on the state and local level. Most environmentalists agreed that Utah was a relatively hopeless place to organize opposition to the project. Utahns tend to be very much prodevelopment, and as a rule do not think very highly of environmentalists (one Utahn called environmentalists "no-growth communists"; another local resident stated, 'I haven't seen a wilderness yet that wasn't improved by building a road through it"). Local opposition to the project in Utah was primarily limited to the Utah Environment Center (UEC). The UEC, the only environmental organization with in-state offices, became involved in the Kaiparowits project in late 1973, when the initial engineering and environmental studies on the project began to appear. The UEC later acted as a clearinghouse for the national groups that mounted a campaign against the project in Utah.

Initially, the UEC made no efforts to change Utahns' perceptions of the plant. However, after the national press conference at Lake Powell, the closet environmentalists in Utah began to offer money and support and the UEC made some efforts in Salt Lake City. The UEC expected to be beaten on the Kaiparowits project, and they felt it would jeopardize their group's future if they came out too strongly against it. Therefore, with the help of national environmental groups, the Kaiparowits Awareness Group (KAG) was formed to spearhead the local opposition to the project and to bear the brunt of the Utahns' animosity.

The KAG organized several rallies on Salt Lake City campuses and held forums to discuss the problems with the project. When the environmental groups finally did venture forth into southern Utah, potential disruptive social effects of the project on the community were stressed, while in northern Utah, the potential adverse environmental effects were pointed to. Only when outside the state did the environmentalists mention possible effects on the national parks, since Utahns were very hostile to this argument. Utahns tend to be ardent believers in states' rights and resist what they consider to be intrusions by outside groups into their affairs.

Generally, the environmentalists came to believe that time was on their side, for even Utahns' perception of the plant became less positive over time. Environmentalists believed that they could delay decisions on the plant by litigation, if necessary. The Sierra Club fought the project by filing a petition with the Public Utilities Commission in California, asserting that it was necessary for a California utility to obtain a permit for building an out-of-state plant (a permit that was not then required).* The Escalante Wilderness Society, the Environmental Defense Fund, and the Sierra Club filed petitions with the Utah State Engineer, attempting to stop him from granting an extension of Utah water rights to the project. An environmental group known as "Save Needed Environmental Levels League," an unincorporated association, filed

* At the time of this writing, the California Public Utilities Commission has recently decreed that out-of-state plants supplying electricity to California must receive such a permit.

a civil complaint in the United States District Court against the Kaiparowits participants, Department of the Interior, Bureau of Land Management and the Environmental Protection Agency charging that such parties were in violation of the National Environmental Policy Act of 1969 and the Endangered Species Act of 1975. Litigation on a variety of other matters was threatened, including a suit against the Secretary of the Interior, charging him with violation of the Congressional mandate to protect the national parks, should the project have been continued by the utilities and approved by the Secretary.

Court suits can provide crucial tools for the environmentalist, bringing publicity and producing delay. There have been criticisms of the ease of access to the judicial system for just these reasons; for example, by L. T. Papay, of Southern California Edison:

> The role of the intervenors appears relatively smooth . . . their role is to question, and in the present framework they can easily present road blocks to large projects in terms of petitions for delays. Such delays, by the financial weight they carry, can sink the mightiest of projects, which leads to the fundamental question: if a utility meets all the laws and regulatory requirements in its applications and submittals, why, then, can it continue to be stopped in its efforts? And stopped, not necessarily by the overwhelming weight of public opinion, but, as we've seen, by the actions of select, elite groups or individuals who, for a simple filing fee, can bring a large multi-billion dollar project to its knees. On the other hand, the energy industries, particularly utilities, must carry the burden of convincing the public of the need for these projects and the efforts being made to minimize their impact. It is not an easy task. We must pay closer attention to public participation and communication for it is the public which should be involved in these issues and not single purpose groups . . . [O]f all the players in the game, only the utilities must consider the welfare of consumers, environmentalists, and others in their decisions [47].

Environmentalists would likely make the following type of response: the political realities are not nearly so simple. There is presently no "public interest" spokesman; there are development interests and environmental interests. Only they, the environmentalists, attribute proper value to stopping environmental decay. Examples such as the loss of clean air in urban areas or the fate of the Great Lakes make environmentalists apprehensive about the ways in which trade-offs have been made in the past and the ability of industry to internalize the real costs of continued pollution. Are utilities really in the position to make proper trade-offs regarding the value of our national parks, they ask, or is the Secretary of the Interior himself, for that matter, if there were not strong environmentalist pressure? For the environmentalists, present standard setting has not proven itself as a means of protecting the environment; many standards are simply too lenient. In addition to the specific emission problems of the Kaiparowits plant, in environmentalist eyes Kaiparowits was representative of a centralized, thermal technology program that would inevitably produce further ecological disruption. Such a policy path may be mutually exclusive with movement toward a policy path stressing conservation and alternative generation technologies. Environmentalists believe that they must fight a plant like Kaiparowits as a means of preventing continued unnecessary pollution and as a way of furthering the transition to an energy path that complements environmental protection.

5
Federal Regulatory Involvement with Kaiparowits

It was the Department of the Interior's (DOI) job to settle the Kaiparowits dispute. The Department's involvement with Kaiparowits raises a plethora of planning dilemmas, some of which this chapter will review.

DOI's environmental review process was initiated in order to comply with the provisions of the National Environmental Protection Act of 1969 (NEPA). NEPA had established certain procedural and informational requirements to be met in the case of all "major federal actions" that could affect the environment. Construction of Kaiparowits would have required a transfer of land from federal to private or state ownership. Also, the Kaiparowits electricity transmission system would have required rights-of-way across federal land. The Kaiparowits land transfer was determined to be such a major federal action, triggering a 28-month impact assessment process.

NEPA was intended to produce a fundamental reordering of federal priorities and objectives regarding the environment, and it has had a great effect upon the day-to-day affairs of federal agencies. DOI's involvement with Kaiparowits provides an excellent example of the attempts of an agency to comply with NEPA provisions and the effects of NEPA requirements on policy. NEPA mandated that agencies balance economic, technical, environmental, and other costs and benefits in their reviews of major federal actions. The Kaiparowits history raises a number of persistent substantive and procedural problems with this "balancing" paradigm. For example, the DOI was to weigh the environmental "costs" of Kaiparowits against its energy and economic "benefits." Given the uncertainty associated with the projection of environmental impacts and the complexities involved in comparing environmental disruption with project benefits, how could such systematic comparison be achieved? Procedurally, was anyone truly in the position to consider fully the trade-offs in this case? The Secretary of the Interior could have terminated Kaiparowits a number of times. Yet the Secretarial mandate remains unclear. As head of the DOI, was his primary responsibility one of protecting the Southwestern parks? Or, given his general policy-making role as a cabinet member, was the Secretary to attempt comprehensive national planning, weighing national, regional, and state impacts? Any resolution of these issues involves significant assumptions regarding the roles of analysis, planning, and authority in energy and environmental decision-making.

This chapter will raise these issues by analyzing the Kaiparowits EIS process within the DOI and setting the context within which Secretary Kleppe would have made his decision on Kaiparowits. The chapter begins first with a discussion of NEPA and the attempt to reorient environmental decision-making.

NEPA

Most of present environmental law has its origins in NEPA, passed by Congress in 1969. Congress recognized that federal decision-making reflected a serious disregard for the environment.* Only a policy requiring a fundamental reordering of the nation's priorities and objectives could arrest and reverse widespread environmental decline. NEPA was intended to ensure that this fundamental reordering would in fact occur within all federal agencies. NEPA established six specific environmental goals for the United States, and directed that the federal government "use all practicable means, consistent with other essential considerations of national policy, to improve and coordinate federal plans, functions, programs, and resources to the end that the nation may" achieve these goals [49]. These goals are as follows:

> (1) [to] fulfill the responsibilities of each generation as trustee of the environment for succeeding generations;
> (2) [to] assure for all Americans safe, healthful, productive, and aesthetically and culturally pleasing surroundings;
> (3) [to] attain the widest range of beneficial uses of the environment without degradation, risk to health or safety, or other undesirable and unintended consequences;
> (4) [to] preserve important historic, cultural, and natural aspects of our national heritage, and maintain, wherever possible, an environment which supports diversity and variety of individual choice;
> (5) [to] achieve a balance between population and resource use which will permit high standards of living and a wide sharing of life's amenities; and
> (6) [to] enhance the quality of renewable resources and approach the maximum attainable recycling of depletable resources [50].

---

* For example, the Senate report recommending passage of NEPA held that "[i]mportant decisions concerning the use and shape of man's future environment continue to be made in small but steady increments which perpetuate rather than avoid the recognized mistakes of previous decades. Today it is clear that we cannot continue in this course. . . . we no longer have the margins for error that we once enjoyed" [48].

The procedural device chosen by Congress to ensure that agency officials would consider environmental variables was the requirement that an EIS accompany "major" federal decisions—those that significantly affect the quality of the human environment.* The ultimate objective of each statement should be to present the final decision-maker with several viable options representing a range of different trade-offs between project benefits and environmental impacts, as well as a range of distinct methods for achieving project goals.

The adequacy of an EIS has become the dominant issue in litigation invoking NEPA. Such litigation has principally focused on the procedural requirements of the act rather than on the act's substantive policy implications. If an impact statement is deemed adequate by the courts and was sufficiently considered by agency decision-makers, then a federal project will not be stopped by the courts under NEPA. As long as a statement has been properly prepared and the balance struck between environmental and other factors has not been "arbitrary and capricious," NEPA requirements have, by and large, been met.

Courts enforce NEPA's impact statement requirement by invalidating agency actions that are accompanied by a defective statement. Courts also examine whether the statement was available before substantial commitments to a project have been made. Violation of the latter criterion indicates that environmental concerns were not a meaningful part of the decision-making process. Projects must be invalidated when an agency's consideration of the statement was clearly not made in good faith.

* NEPA directs the EIS to assess (1) the environmental impact of the proposed action; (2) any adverse environmental effects which cannot be avoided, should the proposal be implemented; (3) alternatives to the proposed action; (4) relationship between local, short-run uses of man's environment and the maintenance and enhancement of long-term productivity; and (5) any irreversible commitments of resources which would be involved in the proposed action, should it be implemented.

Prior to 1973, proponents of projects were responsible for the preparation of an EIS. In 1973, however, the court held that the EIS must be prepared by the responsible government agency. This change in requirement was based on the assumption that the private applicant might frequently overlook or undervalue environmental factors.

Nearly a decade after NEPA was passed, great uncertainty still existed as to how NEPA-mandated changes should be institutionalized. In response to NEPA, agencies have focused on the responsibility for EIS preparation rather than on the substantive goals of the Act. Impact statements are most often written with a judge rather than a decision-maker in mind, complicating the already pressing analytic problem of identifying the trade-offs associated with different decision options, as will be evident in relationship to the Kaiparowits EIS.

EARLY FEDERAL INVOLVEMENT WITH KAIPAROWITS: THE SOUTHWEST ENERGY STUDY (SWES)

Federal involvement with Kaiparowits had begun prior to the preparation of an EIS, as noted in Chapter 2. Secretary of the Interior Rogers Morton had shown early interest in the question of Southwest energy development, and had initiated the Southwest Energy Study (SWES) in May 1971. This decision came in the aftermath of a surge of utility proposals in the Southwest. SWES was to be a detailed study of the entire Colorado River basin, focusing on the region's natural resources and economic needs, as well as on cultural and esthetic factors. The study was to consider the effects of long-term electric power development on both the human and physical aspects of environmental quality and the available alternatives for attaining optimal development with minimal adverse environmental impact.

The SWES concluded that there were no readily available and acceptable alternatives that would allow foregoing Southwestern coal-powered generation in the near future. However, there would be significant environmental effects associated with meeting demand, particularly on air quality. SWES projected that installed generating capacity in the states of California, Nevada, Utah, Wyoming, Colorado, New Mexico, and Arizona would be 56,800 megawatts in 1980 and 109,900 megawatts in 1990. Southern California electricity usage would account for about 60 per cent of this capacity [51]. As a comparison, in 1972 there was approximately 200,000 megawatts of installed generation capacity in the United States. As of 1970, 27,800 megawatts was already being produced in the eight Southwestern states; thus, SWES authors anticipated that a total generation capacity equivalent to 27 plants the size of Kaiparowits would be constructed in the Southwest between 1970 and 1990. SWES has been heavily criticized as being hastily prepared and methodologically simplistic. This latter projection itself makes the study suspect (although the study authors could not have foreseen the demand-reducing effects of the oil crisis and the 1974-1975 recession).

THE KAIPAROWITS EIS

Despite criticisms of SWES, Secretary Morton used it as the basis for his initial rejection of Kaiparowits, the project at this time located at the edge of Lake Powell and designed to produce 5000 megawatts, with only minimal pollution control. Project proponents, however, were successful in convincing White House energy advisors that Morton had made a hasty decision. Thus the EIS process began in January 1974, when Acting Secretary of the Interior John Whitaker designated the Utah office of the Bureau of Land Management (BLM) as the "lead-agency" for EIS preparation.

After an initial organizational period necessary to assemble the EIS preparation "team," BLM began to solicit project information from the utility project manager, SCE. At the outset, however,

SCE was hesitant to release some of the requested information, thinking it unnecessary for BLM's purposes. There was some sentiment in Interior that SCE believed they could push the project through with a minimum provision of information. Interior personnel believed that SCE management perceived any government recommendations as a hindrance to the project. On their part, SCE felt that BLM's requests showed inexperience in preparing power project impact reports and felt compelled to question BLM's desire for increasingly detailed information.

The final EIS (FEIS), over 2500 pages long, was submitted to CEQ on March 4, 1976. The document clearly shows some of the problems with NEPA when it is put into practice. The Kaiparowits EIS did not attempt to rank issues by their importance with respect to the viability of the plant; nor did it include a discussion of the trade-offs between environmental factors and other factors relevant to federal decisions on the project, such as economics or politics. Like most other impact statements, the Kaiparowits EIS was not in a form that would have made it useful as a decision tool. Agencies have tended to compensate for the threat of judicial sanction by producing massive and catalogic reports, hoping to ensure procedural compliance if only by sheer magnitude of environmental detail.

Many critics would claim that the problems with the Kaiparowits EIS are endemic to the EIS process in general. There is widespread consensus that agency compliance with NEPA has been "begrudging and less than complete," as one study puts it [52], although NEPA's goals are generally praised. Many within governmental agencies see EIS preparation as "too costly, time-consuming, inflexible, cumbersome, and detailed" [53]. A number of studies suggest that EISs are used as project justifications, exhibits intended to convince courts that agencies have considered environmental concerns [54] rather than as integral decision-making documents. Thus the environmental impact documentation process may not have much of a role in modifying agency plans, a role that was mandated by NEPA. This partial compliance is likely related both to the nonspecificity of NEPA and to bureaucratic inertia on the part of agencies. For NEPA's goals to be realized, agencies would require a type of creativity that they may not be able to muster.

Bardach and Pugliaresi argue that "Agencies cannot be penetrating or creative when their analyses are directed or mobilized for primarily defensive purposes" [55]. For these authors, the EIS process precludes a "hard look" at environmental impacts for this reason. Like the Kaiparowits EIS, most impact statements do not spotlight the serious issues for the reader. Bardach and Pugliaresi claim that this omission is deliberate: "To highlight the 'critical' environmental issues in a summary section (or elsewhere in the EIS) would necessarily imply that others were not so critical. The solicitor's office (at the Department of the Interior) worries that such judgments are 'conclusory' and in violation of NEPA" [56].

Critics also find a flavor of undue pessimism concerning the environmental impacts of a proposed project in many reports. An EIS must discuss a wide variety of potential ecological developments, even if the magnitudes of their impacts are small. Unfortunately, the qualitative descriptions of these effects often have an alarming ring, although the damages that actually occur may turn out to be quite limited. SCE officials, for example, contend that the continued reference to potential impacts given "worst case" assumptions lead the reader to overestimate probable impacts.* Bardach and Pugliaresi go farther to argue that agencies face legal incentives to make an EIS unduly gloomy:

> Perhaps the most fundamental source of institutionalized pessimism in the EIS is the asymmetry of penalties result-

---

* EIS preparers might reply that they cannot be responsible for the reader's inattention to the fact that "worst case" scenarios are clearly labeled as such and that to try to assign more definite probabilities to ranges of potential impact is often not methodologically possible and would be infinitely more misleading.

> ing from error. "Too little" pessimism can lead to charges of "whitewashing" the proposed development and a court order requiring the department to go back and prepare an "adequate" EIS. No one, however, fears that an EIS will be charged with "tarring" a proposed development and will be held inadequate because it evinces "too much" pessimism [57].

Utilities are very critical of the ways in which federal agencies see the EIS preparation period as a time to develop requirements for mitigating the effects of the project. In the Kaiparowits case, for example, the draft EIS included a proposal in the chapter on "mitigation measures" that Kaiparowits transmission towers should have lights on them so as to minimize the killing of migratory birds. SCE was able to show that such lights would in fact attract birds to the towers. Utility officials felt that they had to spend too much time rebutting similar mitigation proposals, most of which they thought were completely fanciful. To the utilities, it appeared that any proposed mitigation agency would be included in the EIS without any serious review. The utilities believed that it was more rational to discuss possible mitigation measures during the construction permit-granting process that follows the EIS stage.

The most frequently heard criticism of the EIS, however, concerns its length. Most EISs are bulky, often consisting of multiple volumes and thousands of pages. Because early NEPA suits dismissed EISs as insufficiently comprehensive, but never overly lengthy, federal agencies have come to prefer errors of commission to errors of omission. Cynics would claim that sheer size helps to convince courts that an EIS is legally adequate. Overly lengthy impact reports not only have little usefulness as an intraagency decision-making document but also are seriously limited public information sources. Few groups have the time or the resources necessary to review and digest a document the size of the Kaiparowits EIS. It is of more than passing interest to note in this regard that Daniel A. Dreyfus, the Senate staff member who drafted the EIS provisions of NEPA, has indicated that he originally expected federal EISs to be general statements of about two pages in length [58].

Many critics from business and utilities hold that the EIS has merely become a tool for protracted environmentalist litigation. "By declaring the national environmental policy in broad and general terms that invite interpretational dispute," says one critic, "NEPA is fashioned in a manner calculated to breed litigation" [59]. There can be no doubt that a major effect of the EIS requirement has been to give environmental groups a legal and political instrument to cancel, delay, or modify development projects that they oppose.* NEPA has enabled environmental groups to become a significant part of the external environment of federal agencies, thus granting environmentalists an important role in the formation of energy and environmental policy. Professor Keith Caldwell, the so-called "father" of NEPA who worked with Senator Jackson's committee on the legislation, worries that NEPA has become a legal playground, polarizing environmentalists and government agencies instead of providing an avenue for cooperation as intended. It has been argued that protracted EIS legislation will hurt the environmentalists' long-term interests, stereotyping them as self-serving nay-sayers and thereby weakening their creditability.

Whatever its limitations, the EIS process is not universally condemned. EIS proponents hold that the EIS process is well worth the time, money, and effort when all is considered. They hold that the EIS process ensures that agencies take some look at environmental issues and that some look is better than no look at all. Previous to NEPA, the public had little or no access to data nec-

---

* An interesting study by Grunbaum, however, suggests that judges tend to support agencies that are defending projects against environmentalist criticism and also most often use an economic framework to evaluate environmental problems [60].

essary for establishinig coherent positions on environmentally related decisions. The EIS process makes at least some environmental information available, even if it does not present it in the most systematic manner. Bardach and Pugliaresi themselves feel that "it is very likely that many agencies, including the Interior Department, are doing a better job of protecting the environment today than they were . . . before NEPA" [61]. NEPA defenders argue that the problems with the EIS stem from bureaucratic rigidity and legalistic overkill rather than from the objectives of NEPA. As for the charge that NEPA has conferred new power upon environmental groups, supporters hold that environmentalists have catalyzed a widening of the scope of analysis and a broadening of the range of policy alternatives that is considered.

Whether or not the costs of NEPA exceed the benefits depends on the values ascribed to environmental objectives as opposed to other social objectives. Placing high values on environmental goals, however, does not preclude the belief that the present EIS process should be replaced by some alternative. Bardach and Pugliaresi believe that some kind of environmental documentation remains necessary, but that the present process is "anachronistic." They suggest the idea of making agencies relatively immune from litigation on the grounds of alleged defects in an EIS, believing that this step would enable EIS writers to prepare analyses that are "shorter, sharper, and more responsive to the key environmental issues each federal project poses" [62]. Whether such a change could mitigate NEPA's problems and ultimately prove satisfactory is not at all clear.

MAKING TRADE-OFFS: THE PROGRAM DECISION OPTION DOCUMENT

Because most impact statements provide little basis for Secretarial decisions, DOI has developed a structured process to take place between the completion of the EIS and the time for a management decision. The process is to produce a Program Decision Option Document (PDOD), which attempts to make explicit the trade-offs between all of the important factors (environmental, political, economic, technical) involved in the pending decision. The PDOD is usually completed by a staff agency called Program, Development and Budget (PDB) because of its analytical resource capability and lack of constituency (which at least in theory allows for unbiased analysis).

The PDOD requires a great deal of new analysis under pressing time constraints. The Office of Policy Analysis, staffed by about 20 social scientists, had only 30 days in which to prepare the Kaiparowits PDOD. The final product was 49 single-spaced pages, organized into two parts. Part 1 focused on the major issues involved with approval, rejection, or delay of Kaiparowits. It summarized the environmental data from the EIS, and included a brief comparison of the cost of Kaiparowits with the costs of a nuclear, combined cycle, or coal plant sited in California (the California coal or nuclear plant was indicated to be less costly than Kaiparowits). Part 1 then summarized the arguments for project approval (viz., production of electricity, displacement of oil, stimulation of southern Utah's economy), rejection (viz., air quality impacts, electricity demand uncertainty), and delay (viz., allow for a regional EIS, an independent study of electricity demand, and resolution of the Clean Air Act amendments). It also indicate the "positions of interested government parties" on the project, those parties being EPA, FEA, the Energy Research and Development Administration (ERDA), and officials of the states of Utah, California, Arizona, and Nevada. Part II of the PDOD reviewed the alternatives for the various project components (that is, the plant site, townsite, limestone quarry site, and location of the transmission lines) should the project be approved.

After a PDOD is prepared, it is circulated among the Assistant Secretaries for a recommended decision on the project and a short justification of their recommendation. The Secretary's staff reduces these recommendations into a decision memo that is presented to the Secretary along with other documentation. At least 72 hours after the Secretary receives this information, a meeting between all interested Interior parties is held, at which time the Secretary asks questions and listens to arguments for one or two hours. This meeting provides the final internal input into the Secretarial decision.

This system of decision making was not heavily used prior to Secretary Kleppe's stay at Interior. Kleppe came to believe that the due process mechanisms of this system were very important. In Interior, unlike many other federal departments, there are strong bureaus with different constituencies, allowing for considerable in-house debate. Ideally, this debate should allow all major considerations to surface.

There are, however, some apparent limitations of this process. First, most of the Department bureaus see themselves as providing a service to a certain constituency or set of interests. For example, the BLM has traditionally been concerned with resource development of the West, while the National Park Service (NPS) seeks to protect its parks. Thus, these bureaus tend to push certain values rather than to seek "objective" analysis of policy trade-offs. Second, the relative economic viability of alternatives or incremental project changes are not considered until the PDOD stage, when time is running short. Therefore, the leverage that economic considerations can have may be limited. Economic analysis is not usually completed for the EIS because many DOI officials believe that the EIS should present only environmental data; otherwise, they argue, the report might become an attempted justification of a prechosen alternative. Third, the Secretary tends to be insulated from the decision process until a final decision is pending. His role may therefore be limited to making a yes-no-delay decision rather than having input at a much earlier stage when he could affect project design without the consequences of requiring a new EIS and bringing substantial delay. Fourth, and conversely, the decision-making system forces resolution of extremely technical questions at the level least prepared to deal with them, the high management level. For example, a final decision on Kaiparowits would have involved over 40 separate decisions about such project details as transmission corridors, limestone quarry siting, etc.

Ultimately the decision process in a case such as Kaiparowits culminates in a set of value judgments on the part of the Secretary. As discussed in Chapter 3, there was uncertainty over the extent of Kaiparowits' environmental impact, and there was even debate over the need for the project. The PDOD noted these uncertainties, and compared Kaiparowits economics to other generating sources, but such comparison gives the Secretary no calculus with which to make his decision. The problem is an obvious but important one: potential environmental disruption cannot be easily weighed against the production of more energy; to do so would require an *a priori* establishment of the value of avoiding a given amount of pollution against the value of producing a given amount of energy. Any indexing is value-laden and highly debatable.

To say the Secretary would have had to make a value judgment is only to say that the analytic process could only clarify, and to a limited extent, quantify the issues. The Secretary would have had to decide if protecting the parks, for example, merited special attention because of his role as head of Interior. He would have had to decide what alternatives to the project that he would and would not consider, such as conservation potential in California. He would have had to determine the levels of risk that were appropriate for considering decisions affecting the canyonlands of the Southwest. He would have had to decide whether a decision on Kaiparowits could be made without ordering a regional EIS on the effects of all proposed Southwest energy development. To determine the final issues upon which the fate of Kaiparowits would have been decided, it is important to consider the in-house conflict and external pressures that Secretary Kleppe was facing.

CONTEXT FOR KLEPPE'S DECISION

At the Bureau level within the DOI, there was considerable infighting. The Park Service was against Kaiparowits from early on, Energy & Minerals was for the project, and Land & Water was indecided. Land & Water was undertaking extensive staff work in order to come up with a well-considered recommendation. Outside interest groups, realizing the uncommitted posture of Land & Water and recognizing a potential ally with access to the primary decision-maker, focused some of their efforts there. Both environmentalists and utility representatives had a number of meetings with the division. The Park Service, on the other hand, had very little contact with outside groups because their position was already determined. They instead concentrated on the preparation of their study of Kaiparowits air quality effects, which predicted very significant adverse impacts (see Chapter 3).*

Secretary Kleppe saw two possible perspectives on the Kaiparowits project as the DOI decision-making process neared its conclusion—either the Kaiparowits plateau coal was valuable enough to mine in the light of project development consequences or it was not [63]. Kleppe was not persuaded by California Energy Commissioner Ronald Doctor's argument that Kaiparowits would

---

* There is some feeling that the timing of the study was an important political play on the part of NPS, which strongly desired to have the Bryce Canyon and Glen Canyon National Parks redesignated to Class I air quality levels, assuming that such a designation would stop Kaiparowits. NPS contended that there was no underlying strategy to their study; it was needed because they believed the EIS to be inadequate.

be a bad investment for the utilities. Kleppe believed that the private sector could take care of itself regarding its financial decisions.

Kleppe has stated that he would have made his decision with regard to Kaiparowits alone; he would not have considered the possibility of cumulative effects from the operation of a number of coal plants in the region. The Secretary also refused to consider the issue of redesignation of Southwest national parks to Class 1 areas; such redesignation would have been within his discretion. Kleppe felt, however, that Congress should resolve the redesignation issue in its Clean Air Act amendment deliberations Kleppe also rejected the options of delaying his decision or requiring an entire new EIS to consider alternative plans, believing that further delay was undesirable.

Any decision on the controversial project, whether positive or negative, would have entailed a good deal of political risk for Secretary Kleppe. Upon his approval, the future of the project would have still been up in the air. Further environmentalist litigation was inevitable—litigation that could have conceivably stopped the project. Also, some DOI officials were coming to believe that the project would never get off the ground for economic reasons. If the Secretary had approved the project and it subsequently was cancelled, he would have needlessly alienated environmentalists.

A negative decision could have also proven to be politically damaging. A pending Nuclear Initiative in California, if passed, would have made the siting of nuclear plants in California nearly impossible in the short to medium term. If the Secretary denied Kaiparowits applications and subsequently the Initiative was successful, California's energy future would have been extremely uncertain and Kleppe would have been partially responsible.

Secretary Kleppe was never forced to make a decision on Kaiparowits, and he has said that he does not know what he would have decided. He did add that he would have disregarded the political pressures involved. There are new proposals for coal plants in the Southwest on the books,* yet the decision-making issues and difficult value trade-offs raised by Kaiparowits remain unresolved.

* Intermountain Power Project (IPP), the 3000-MW so-called "son-of-Kaiparowits," has been proposed by a consortium of Utah utilities backed by the Los Angeles Department of Water and Power. The original site was 70 miles northwest of Four Mile Bench; the site was undefined as of this writing.

6

Economics and Finance*

Between 1963, when Kaiparowits was first proposed, and 1976, when the plant was finally cancelled, the world of electric utilities changed enormously. Previous chapters have described how, during this period, environmental politics and regulations became formidable obstacles to new power plant development. While this occurrence played a vital role in the demise of Kaiparowits, it would be wrong to pinpoint it as the only cause of the power plant's cancellation. The financial requirements of electric utilities and the economics of power plant construction also changed radically in the 13-year period before 1976. Though the exact impact of these changes on the cancellation decision can only

* Much of the information in this chapter is based on data provided by, and interviews with, the staff of Southern California Edison (SCE). Although Southern California Edison was not the only participant in the project, it was the largest and was also the project manager.

be speculated upon, they are likely to have played a primary, if not decisive, role.

In late 1963 a large coal-fired electric generating plant on the Kaiparowits Plateau in southern Utah appeared to be an economic godsend. Demand for electricity, the fundamental reason behind building any power plant, was growing with no decline in sight. The coal resources underneath the plateau could easily fuel the power plant for its lifetime. The idea of a fuel supply reserved for the use of the power plant, and controlled directly by the utilities, was very appealing. This rosy picture continued throughout the 1960s.

Three economic and financial developments of the 1970s diminished the attractiveness of Kaiparowits. First, estimated costs of the project skyrocketed during the project's planning life; forecasted per-kilowatt costs of the project grew fivefold in 13 years (in escalated dollars—see following paragraph). Second, expected growth rates of electricity demand fell significantly after the OPEC oil price increase of 1973 and the economic recession of 1974-1975. Third, the financial status of electric utilities declined drastically in the mid-1970s, causing the financing of new projects to become increasingly problematic. This chapter will examine each of these factors in detail.

Following the procedures of SCE, the "escalated dollars" used in estimates are not in constant (uninflated) terms. Instead, they are a summation of the estimated *nominal* dollars expected to be spent at differing times during construction. These estimates reflect assumptions regarding both real cost change and inflation. Even with perfect economic information, different year dollars are not directly comparable. A price index of some sort is necessary to put different year actual dollars into common (that is, comparable) dollars. Because SCE data reflect *anticipated* dollars, a price index cannot be used to translate different year escalated dollars into a common unit. Thus, accurate comparisons of cost estimates at different times are not possible, and the available information thus provides a somewhat limited perspective.

COSTS

During Kaiparowits' 13-year planning period, the estimated per-kilowatt (kW) cost of construction multiplied fivefold, and the estimated mining cost per ton of the associated coal deposits increased tenfold (in escalated dollars). The reasons behind these spiralling costs shed light not only on the cancellation of Kaiparowits, but on a variety of other energy issues facing the U.S. in the mid-1970s.

It is difficult to compare cost estimates that were prepared as many as 13 years apart. The plant design, which provided the base for the estimates, changed many times during the period. To mitigate this difficulty, this chapter presents cost estimates on a per-kilowatt basis. Table I summarizes the estimated capital costs as Kaiparowits planning progressed. The upward progression of costs is dramatically apparent. While the dramatic change in estimates over the entire planning period is interesting, the estimates prior to 1971 may be somewhat misleading because of the differences in plant design and detail of estimation. The 140 per cent increase in estimated cost from 1971-1976 is still quite significant since plant design remained relatively constant during this period. It would be wrong, however, to ascribe all or most of this increase to factors pertaining to Kaiparowits alone. There were five basic factors underlying the increases in estimated costs: economy-wide inflation, real-cost escalation in the construction industry, changes in regulatory requirements, changes in estimation procedures and detail, and changes in plant design and schedule.

Almost all types of heavy construction experienced rapid price escalation between 1963 and 1976, with a particularly sharp burst between 1973 and 1975.* Table II shows three different price in-

* A November 1975 article in *Fortune* cited the example of a chemical plant that was started in 1974. A year later, after an identical addition was authorized for the same location, the estimated cost had risen by 50 per cent.

Table I. Estimated Capital Costs[a]

Plan Date	Total Capacity (MW)	SCE Share	In-Service Dates	Generation ($/kW)[b]	SCE Total ($/kW)[c]
12/64	5000	100%	1972-1981	100	220[d]
7/70	2000	60	1979-1980		250
8/71	2000	50	1978-1979	315	492
9/72	3000	44	1979-1981	493	649
6/73	3000	40	1980-1982	477	627
12/74	3000	40	1981-1983	628	846
1/75	3000	40	1981-1983	876	1102
7/75	3000	40	1981-1983	876	1138
2/76	3000	40	1982-1984	918	1184

[a] The capital cost estimates used represent the sum of accounting costs expected to be incurred prior to the time the plant begins operation. The amounts are summed without discounting for the time value of money and without reducing denominations to common year dollars. This is equivalent to an estimate of the book value of the plant at the time it is placed in operation, using historical cost accounting assumptions.

[b] These numbers represent the costs of the generating station. Omitted are the costs associated with the transmission systems, which varied for each of the project participants.

[c] These numbers represent SCE's total costs, including its transmission lines to Kaiparowits.

[d] Documents provided by SCE show that $220/kW was the best estimate of total SCE cost for this period; however, there are many indications that $100/kW figure was also used at that time. As late as April, 1976, William Gould, of SCE, was quoted in *Electric Perspectives* using a conceptual estimated cost of $500 million for a 5000 megawatt plant, which equals $100 per kilowatt. It is unclear, however, whether this figure referred to the generation cost or the total cost.

dexes to give some perspective on how Kaiparowits fared relative to price changes for other goods and services. The first column shows general economy-wide inflation, while the other two columns depict price level changes in heavy construction. These indexes represent the actual price levels in effect at the date given, while the Kaiparowits estimates for the same dates (shown in Table I) are in expected future costs. A change in construction costs, as reflected in the price indexes, does not necessarily cause a change in SCE's estimate since these estimates include an expected price escalation factor. SCE would revise their estimates

Table II. Selected Price Indexes (1972 = 100)

Date	GNP Total (All)[a]	GNP[b] (Non-Residential Structures)	Electric Light & Power Plant[c]
1964	73	64	67
1970	91	88	88
1971	96	94	94
1972	100	100	100
1973	106	108	107
1974 (4th Qtr.)	121	138	140
1975 (2nd Qtr.)	125	141	150
1975 (4th Qtr.)	129	143	153

[a] Implicit price deflator for gross national product: Total (from Economic Report of the President, January 1976).

[b] Implicit price deflator for gross national product: Gross Private Domestic Fixed Investment in Non-Residential Structures (from Economic Report of the President, January 1976).

[c] Handy-Whitman index based on costs of the inputs used in constructing electric light and power plants and utility buildings. Consists of weighted averages of labor, materials, and mechanical and electrical costs. Costs are based on information from the Bureau of Labor Statistics, industry groups, labor unions, and reports in publications such as *Engineering News-Record*. Labor, material, and equipment costs are aggregated by their 1950 weights.

Source: Bureau of the Census, Construction Reports C-30, Value of New Construction Put in Place.

if the actual escalation experienced after an estimate was different than that expected. Additional revisions of estimates would be required if SCE changed their expectations regarding future escalation rates.

In part, the changes in estimation can be attributed to misjudgment in estimation, rather than to actual cost increases. When a plant is first proposed, there is very little specific information available about it. Many of its characteristics remain unknown, and cost estimates can only be based on the costs of similar plants for which information is available. Only as preconstruction engineering studies are performed does detailed information about plant specifications become known, allowing the utility to make more accurate estimates.

Seemingly trivial changes in estimation procedures can produce a significant change in forecasted costs. In 1974, SCE changed its assumed price escalation rate from simple to compound interest. This change alone caused an addition of $105 million, $35/kW, to the estimate of total plant costs. SCE also did not anticipate the magnitude of the cost escalation that was to occur during 1974. The previous estimate had anticipated an annual escalation rate of 8.1 per cent for the labor and materials needed for Kaiparowits. As of September 1, 1974, the average actual annual increase for 1974 was over 19.7 per cent, necessitating a $115 million increase in the estimated base cost of the plant. These examples indicate how changes in estimation procedures and unanticipated changes in the economy can cause cost estimates made at different times to be significantly different, even if the real cost of building the plant had not changed. ("Real" in this sense means the cost of the plant if built as planned, if looked at after the fact so that there would be no estimation uncertainty.)

The most controversial cause of Kaiparowits' cost escalation, and the factor most relevant to public policy, stemmed from the requirements imposed upon the plant because of new environmental regulations. According to SCE staff estimates, by the time the project was cancelled, environmental regulations had added $641 million to the cost of the project. This amounted to $214/kW, or about 25 per cent of the 1975 estimated total generating plant costs. Some of this cost increase was the direct result of requirements such as the addition of scrubbers to meet air pollution standards. Other costs were incurred indirectly. For example, the initial site was not acceptable and the alternative site required different equipment, such as a 20 per cent larger boiler than that needed at the earlier planned site.

The $641 million estimate of escalation due to environmental requirements reflected the following costs:

- $300 million for scrubbers
- $190 million for precipitators
- $ 40 million for additional water pipeline
- $111 million for environmental studies, waste treatment facilities, and different equipment required because of the higher altitude of the new site.

It is unlikely that the $641 million figure fully captures environmental costs. A crucial but difficult to quantify cost of environmental considerations stems from the resulting delays in plant construction. There are two basic ways in which delays caused by environmental and other factors increase costs. First, because of inflation, postponing an expenditure usually means that it will cost more than originally planned. It is clear that this effect was occurring in the Kaiparowits case, but it is difficult to isolate its magnitude because of the changing specifications and estimation misjudgments, as discussed above. Second, environmental regulation or opposition can cause the amount of construction time to be different than planned. If the construction time is lengthened, the early expenditures will be tied up longer in the project, which is not yet producing revenue. This results in greater financing costs, as well as higher tax and insurance costs.

A further potential cost of environmental regulation stems from the possibility that even if the plant were built, regulations might prevent it from being permitted to operate. It is not certain that the Public Utilities Commission would allow the utilities to recover the costs of building the plants. If it did not, the entire cost would be borne by the utilities' shareholders, not the ratepayers.

While estimates of the costs of constructing the Kaiparowits plant were rising, estimated fuel costs followed suit. Table III shows how SCE's estimated 1982 Kaiparowits coal price changed over time.

A fundamental motivation behind building the plant at Kaiparowits was to take advantage of the coal deposits found there. The

Table III. Projected Kaiparowits Coal Costs for the Year 1982

Date of Estimate	Cents/Million Btu	$/Ton[a]
5/71	19	4.10
8/72	32	6.91
5/73	39	8.42
9/74	84	18.14
4/75	140	30.24
2/76	180	38.88

[a] Based on an average of 10,800 Btu/pound.

proximity of the plant to its fuel source was a major economic advantage. If not for the vast coal deposits there, Kaiparowits would have remained an obscure desert plateau. But the advantages of cheap fuel clearly decreased in the last years of the proposal's life. A July 1974 estimate of operating and capital costs showed fuel as 19 per cent of the total cost of Kaiparowits power. A similar estimate in July 1975 listed fuel as 33 per cent of the total costs. Not only was the estimated cost of Kaiparowits coal going up rapidly, but its relative importance to the delivered cost of electricity was rising even faster.

The reasons for the rapid rise in coal cost estimates are difficult to identify. Clearly these estimates rose as they were adjusted to include the costs of constructing a new deep mine, developing new mining techniques, and training local Utah citizens to become miners. Many of the factors affecting construction of the power plant may also have affected the estimated cost of coal. Changes in productivity caused by the passage of the Coal Mine Health and Safety Act of 1969 have also been cited as a main cause of the rise in expected costs. This explanation does not seem to fit the facts of the Kaiparowits case. Most of the regulations associated with the Act were known by the end of 1970. Productivity in Utah mines was actually higher in 1974 than in 1970, despite the fact that the national average was lower than Utah's in 1970 and had declined steadily through 1974.

Whatever the exact reasons for the escalation in Kaiparowits generation and fuel costs, they caused the plant to become a less cost-effective proposition. An April 1976 Stanford Western Energy Policy Study (SWEPS) report examined the economics of building Kaiparowits compared to other potential plants. The report used cost estimates submitted by SCE to the California Energy Commission in December 1975. Based on those estimates, the analysis found that the economic advantage of Kaiparowits was extremely sensitive to the future costs of fuel and the real discount rate used. (The real discount rate equals the nominal discount rate minus the rate of inflation. For example, if a ten per cent discount rate is used during a period of six per cent inflation, the real discount rate is four per cent.) If the delivered cost of oil went down to $10/bbl in 1975 constant dollars (the pre-1973 price was less than $6/bbl), and the cost of coal went up to $30/ton, a combined cycle oil plant would be more economical than Kaiparowits, regardless of the discount rate. The SWEPS analysis concluded:

> The net economic benefit of displacing oil used in existing plants with electricity from Kaiparowits depends strongly on the real discount rate. At a 4 per cent real discount rate, operation of Kaiparowits creates a cost savings, but at an 8 per cent rate, Kaiparowits creates a loss.

Unfortunately, it is difficult to discern SCE's real discount rate. In SCE's submission to the California Energy Commission, they stated that 13 per cent was the nominal discount rate used for internal economic evaluations, but the assumed rate of inflation was not stated. Therefore, the SCE real discount rate number needed to draw a conclusion from the SWEPS report cannot be deduced. Nevertheless, the SWEPS analysis does show that the merits of Kaiparowits were highly sensitive to a number of parameters that could not be predicted with certainty.

DEMAND SHORTFALL

Between 1973 and 1975, the historic growth of electricity demand came to an abrupt halt. This changing level of demand for electricity was discussed in Chapter 3. Table IV below shows SCE's demand forecasts as SCE attempted to follow this radically changing situation.

The shortfall between actual 1975 peak demand compared to the 1973 forecast of 1975 peak demand is almost twice as large as the capacity of the Kaiparowits plant. The difference between the 1975 and 1973 forecast of 1985 demand is almost five times the size of Kaiparowits. A short-run effect of the demand decline, surely weighing on the minds of utility executives in 1975-1976, was a considerable oversupply of generating capacity. SCE's reserve margin* in 1975 was 34 per cent, much higher than SCE's

Table IV. SCE Demand

	1973 Forecast (MW)	1975 Forecast (MW)	Difference (MW)	% Below Forecast
1975 Peak Demand	12,217	10,193 (actual)	2,024	17
1980 Peak Demand	16,683	13,471	3,212	24
1985 Peak Demand	22,123	17,081	5,042	23
1975-80 Peak Demand Compound Annual Growth Rate	6.4%	4.7%		
1980-85 Peak Demand Compound Annual Growth Rate	5.8%	4.9%		

Source: Estimates provided by SCE to FEA Final EIS, p. I-45. Actual 1975 figures from 1975 SCE Statistical Report.

* See footnote, p. xx.

practice of maintaining a reserve margin of 18 per cent, plus or minus 2 per cent.

An effect that cannot be measured by the change in demand forecasts is the uncertainty about future demand that resulted from this changing environment. Previously, utilities were able to forecast demand fairly accurately; now they were much less sure of their predictions. Economists in 1976 could not know how much of the recent fall-off in consumption could be regarded as a one-time nonduplicable response to appeals for energy conservation, and how much could be attributed to price elasticity (about which, for electricity, relatively little was known). Obviously, the more that price-elastic responses to rising prices played a part, the more future growth was apt to be reduced. Not having a fix on this question, forecasters were operating in the dark. Clearly, however, the need for the power from Kaiparowits was not completely certain.

Another uncertainty facing SCE at the time the Kaiparowits plant was cancelled was who would use the 18.6 per cent of the power output that remained untaken. Until the Salt River Project (SRP) dropped out of the Kaiparowits plant in May 1975, the participation in the project was as follows: 40 per cent SCE, 23.4 per cent SDG&E, 18 per cent APS, 10 per cent SRP, and 8.6 per cent unsubscribed. After SRP dropped out, the remaining participants were obligated to assume the 18.6 per cent of the power that was unsubscribed unless other participants could be found. SCE has stated that a number of other utilities had expressed interest in the unsubscribed power, but that the participating utilities were holding back before selling it in order to determine what its real cost would be and to reserve the power for a Utah utility in case one should desire it.

In any case, the power remained untaken when the plant was cancelled in April 1976. The unsubscribed portion of the project represented a potential cost of over $500 million. If each of the remaining participants was forced to assume a proportionate share of this cost, SCE's share would be over $250 million, and it would need to absorb an additional 275 megawatts of capacity into its system. This possibility almost certainly compounded the effects of demand uncertainties.

SCE staff claim that neither the decrease in forecasted demand nor Kaiparowits' subscription problems killed the plant. They also hold that since the cost of oil quadrupled since 1972, it would be economic to build a new nonoil plant such as Kaiparowits and substitute it for an existing oil- or gas-burning plant. Also, new plants would be needed, even with lower expected demand, and Kaiparowits would have been more economic than other potential plants.

Though the exact impacts of demand and shortfalls and uncertainties on the decision to cancel Kaiparowits cannot be determined, they certainly raised questions about the necessity of the proposed plant that were not previously anticipated.

FINANCE

Before 1973, as utilities invested increasing sums to build new plants and equipment, the size of their rate base expanded and regulators allowed them to earn a higher rate of return on this investment. Growth in sales almost automatically provided growth in earnings, without the need for higher rates.

In 1973, the picture changed radically. Oil price increases plus general inflation cut into utility profitability by increasing costs before regulatory bodies could increase revenues. Even if growth were to surge forward once again, growth alone could no longer guarantee adequate earnings. Inflation and fuel price rises had made it necessary for utilities to petition state regulators time and time again for rate increases, after years in which electricity rates had actually declined. Regulators, besieged by irate consumers, were reluctant to raise rates. By 1976, when Kaiparowits was cancelled, few utilities could operate with the absolute assurance that they would be allowed revenues sufficient to cover the costs of building new plants to meet demand.

The sponsors of Kaiparowits were not immune to these trends. These factors led to serious declines in stock values. The price of common equity shares of each of the project's sponsors is traced in Table V.

To raise money for new plants, utilities must eventually issue new shares of common stock. Raising money during a period of low stock prices forces firms to issue proportionately more shares than in periods of higher stock prices. Each of these shares is entitled to dividends just like any other share.

Investors in utility stocks appear to place considerable emphasis on stable dividends. As a consequence, utility executives do not consider it financially sound to decrease dividends per share; in-

Table V. Common Stock Share Price at December 31 ($)

Year	SCE	SDG&E	APS
1967	36.75	21.00[a]	23.38
1968	36.13	24.50	25.63
1969	30.13	22.75	21.38
1970	32.13	23.00	22.00
1971	29.50	21.00	20.25
1972	27.75	20.25	24.00
1973	18.50	14.75	18.13
1974	17.50	10.13	12.00
1975	19.63	11.00	15.75
1976	22.83	14.63	19.63

[a] Adjusted for 2-for-1 stock split in May 1968.

deed, utility managers feel pressure to raise dividends per share periodically. Under these circumstances, total dollars paid out by a utility as dividends are a function of the number of shares outstanding, not the proceeds of the issues of new shares. In a period of depressed stock prices, utilities can issue new stock shares only at the expense of much higher dividend costs in the future. Ironically, this process can become self-reinforcing. Higher dividend costs in the future mean that the utility will need to pay a higher proportion of its future net income as dividends. Consequently, it can rely less on internal funds in the future and will be forced to sell more shares, which will in turn produce subsequent problems in meeting dividend requirements.

A second problem accompanying the sale of new utility shares during a period of low stock prices is the loss of wealth to current shareholders. This loss occurs through a phenomenon called dilution, which requires explanation.

Textbook finance conceives of stock prices as an indication of the return an investor expects on his investment. For example, if the equilibrium risk-free rate of return were 20 per cent, a riskless investment yielding $20 per year in perpetuity would cost $100.

Public utility commissions usually determine a utility's allowed profits on the basis of its rate base (book value) per share. A firm's book value is the sum of its retained earnings and the historic values of its previous equity issues. Regulators multiply a utility's book value by a "reasonable rate of return" to determine allowed profits, and then ostensibly set prices at a level high enough to generate the targeted profit level.*

Suppose a utility's book value per share is $100 and the public utility commission allows a rate-of-return on book value of only 10 per cent. Then earnings per share would be $10, and the market price of each share will fall below book value per share until

* It will be assumed for the purposes of this chapter that the utility's book value equals the value of its rate base and the value of construction work in progress. In actuality, the deviation between the sums of these values and book value is less than a few percentage points. Without this simplifying assumption, this section would become overly detailed and unnecessarily technical. The simplification does not affect the conclusions of this section.

the $10 yields the previously assumed 20 per cent market rate of return. Each share will then sell for $50, despite the $100 book value per share.

The issuance of new stock when existing stock prices are below book value causes dilution, which is a disproportionate reduction of previous shareholders' percentage ownership of the firm. This process can best be shown by extending the example above. Suppose that, at a given time, there are ten shares of stock in the utility, each with book value per share of $100 and market value of $50. The firm's total book value will be $1000, its total market value $500. If the company needs to raise $50 to build a new plant, it will need to issue a new share. This new share will have a claim on company assets and income equal to the previous shares. Yet since the new share will sell for an original price of $50, it will contribute only $50 to the book value of the firm. After issuing the new shares, the utility will have a total book value of $1050, the sum of the book values of the old and new shares.

Whereas before the new stock sale the firm earned $100 on a $1000 book value, now it will earn $105 on $1050 book value. But whereas the older shareholders received 1/10 of $100, or $10 per share, now they will only receive 1/11 of $105, or $9.50 per share. As a consequence of the new stock sale, owners of old shares have had their portion of utility profits "diluted." If building new plants requires new stock sales, existing shareholders will want to avoid such expansion when the utility's market value per share is below the book value per share.

At the end of 1975, each of Kaiparowits' sponsors was in a situation where new stock sales would cause dilution. Table VI compares each utility's market value per share to its book value per share.

Utility executives may fear dilution for reasons other than concern for existing shareholders. The prospect of dilution may completely eliminate the option of new stock sales. Companies that must constantly sell new stock cannot easily do so with a record of falling earnings and a reduced potential for dividend payments. But the implications of continuing stock offerings below the firm's average book value per share go far beyond this effect. The first such offering hurts only the company's old stockholders. But the investor who buys stock in the first sale becomes the victim of the second, and the third, and so on. Eventually it is to be expected that investors will catch on to what is happening. If they are acting rationally, they will simply refuse to buy the company's stock at all. At that point, utilities may have lost access to the capital markets, an occurrence that is not easily reversed. "Utility investment reputations," says one utility executive, "are like virginity; they can be preserved but not restored."

Table VI. Market/Book Ratio[a] at December 31

Year	SCE	SDG&E	APS
1967	1.76	1.58	1.48
1968	1.62	1.77	1.59
1969	1.26	1.55	1.27
1970	1.29	1.49	1.26
1971	1.12	1.30	1.09
1972	1.01	1.17	1.20
1973	0.64	0.87	0.87
1974	0.58	0.60	0.60
1975	0.63	0.68	0.79
1976	0.71	0.88	0.95

[a] $\dfrac{\text{Common Stock Share Price}}{\text{Total Equity} \div \text{No. Common Shares Outstanding at Year End}}$

It is impossible to determine precisely the degree to which fear of dilution encouraged the cancellation of Kaiparowits. But it is clear that the effect was to make oil plants an increasingly attractive alternative for meeting future power demand relative to coal plants. An oil plant and coal plant with identical generation capabilities differ in the amount of capital investment needed to build them. An oil plant burns expensive fuel but requires a relatively small capital investment; a coal plant requires a large capital outlay but incurs relatively low fuel costs. When new stock offerings entail dilution, utilities face incentives to minimize financial requirements by building power plants that require the least capital. As a consequence of low stock prices, Kaiparowits became less attractive relative to oil plants, which were less capital intensive. The same factors were affecting utilities' new debt issues, many of which were down-rated during 1974-1975. This bond down-rating resulted in higher interest costs and greater difficulties in raising debt capital, further lessening the attractiveness of capital-intensive plants.

One additional financial constraint on Kaiparowits was the impact of a regulatory device known as the allowance for funds used during construction (AFDC). When a plant is built, it is usually under construction for a number of years, particularly in the case of capital-intensive plants. During that time, the utility has invested money and desires a return on that money. However, the customers using the utility's product at that time do not benefit from investment in the incomplete plant. Instead, the customers who use the product at the time that plant is in operation will benefit from the investment. For this reason, the Public Utilities Commission in California and most other commissions have allowed the utility to earn a paper (not cash) return on this investment during construction. That is, a return on investment is recorded during the period of construction but is not received in cash until the plant is actually in operation. As a result, financial analysts have come to look not only at the level of utility profits but also their "quality." Earnings that in large part consist of AFDC paper profits are valued less highly. The utility's investment thus yields a less highly valued return. Table VII shows the magnitude of this effect on Kaiparowits' sponsors during the period 1972 through 1976.

Construction lead times, the number of years from the commencement of construction until the plant begins operation, in-

Table VII. AFDC as Per Cent of Profits after Tax[a]

Year	SCE	SDG&E	APS
1972	5.2	9.2	15.0
1973	6.9	14.3	19.9
1974	7.4	11.1	32.2
1975	14.3	31.9	28.7
1976	21.5	27.2	35.5

[a] Before preferred dividends.

creased in the 1970s from the average prevalent in the 1960s. The longer a plant's lead time, the longer its contribution to earnings is withheld as AFDC. In conjunction with increases in capital expenditures, these lead times meant that a growing portion of utility profits was tied up in paper AFDC. While companies would eventually receive these monies, they were not of current use. Instead, firms needed to obtain external funds, adding to their already growing burden of external financing. Because capital-intensive plants tie up large amounts of earnings in AFDC, this requirement provides an incentive against building capital-intensive plants, particularly plants with long lead times. Kaiparowits was a capital-intensive plant and subject to a very long lead time. Both of these characteristics tended to diminish Kaiparowits' attractiveness as an investment.

CONCLUSIONS

In early 1976, when the decision to cancel Kaiparowits was made, decision-makers at SCE faced a great deal of economic uncertainty. The need for power from Kaiparowits was uncertain; past demand trends were no longer applicable and future trends were not yet discernible. Estimates of the cost of constructing the plant were rising and each estimate cast doubt on the one before it. Estimated fuel costs had inexplicably skyrocketed, and no one could predict when they would level off. Finally, stock prices had plummeted to well below their book values, and selling new shares to finance Kaiparowits would compromise the interests of existing shareholders. All of these factors made Kaiparowits a far less attractive investment prospect than when it was first proposed 13 years earlier.

7
Conclusion

When the participating utilities pulled Kaiparowits out of their planning schedules, they were extricating themselves from a regulatory situation that SCE Project Manager Robert Currie called "the most complicated mess I've ever seen." In addition to the economic and demand uncertainty reviewed in the previous chapter, uncertainty partially brought on by environmentalist-induced delay, the utilities faced the prospect of applying for more than 220 permits and authorizations from 42 federal, state, and local agencies—*if* Secretary Kleppe had approved the project. The inevitability of increasing environmentalist litigation and the pending Clean Air Act amendments heightened the uncertainty over the project's fate. Along with the fact that the cost of the project was continuing to increase, it was no wonder that Kaiparowits was cancelled. A collision between environmental and development increases was thus averted: the Kaiparowits issues did not have to be faced.

Some utility officials have conceded that they are now "gunshy" because of Kaiparowits. They believe that it is "nearly impossible to get such a large installation off the drawing board any more." Many project supporters feel that the demise of Kaiparowits was only partially a result of its proposed location. Even if it would have been sited elsewhere, they maintain, it would have been opposed by groups that are against energy expansion and could have been thwarted by arbitrary, bureaucratic regulatory politics.

Before the 1970s, the rules of the game were much simpler. Demand predictions were quite accurate, and the main task of system planning was to schedule enough new plants to meet that demand. Financial or regulatory constraints, if they were a factor, would usually result in a change in the timing of construction projects; cancellation was rare. Pre-1970 power plant siting in the state of California, for example, was fairly informal, with "friendly agreements" reached on plant sites between the resources agency, conservationists, and the utilities.

Many would say that Robert Currie's aforementioned comment well characterizes the present state-of-the-art for energy and environmental decisions. Kaiparowits history raises a number of social and policy questions that lead rather far afield from the Kaiparowits plateau. Previous chapters have discussed the technical uncertainty involved with projecting the physical consequences of energy projects: analysts can assign probabilities to a range of possible impacts, but the margins of error remain quite large. All parties agree that the combustion of coal for the production of electricity has some deleterious environmental results; the debate regards how well such problems can be mitigated and, more importantly, what constitutes unacceptable environmental damage. Neither NEPA, which mandates only procedures, nor pollution standards, which set "acceptance levels," give us explicit theories for making trade-offs or for valuating the environment. Furthermore, there is disagreement over whether the market system can properly internalize environmental costs even if there were consensus on how it should be valued.

Decision-makers are thus presently faced with what Harvard Law Professor Lawrence Tribe calls "the universal difficulty of choosing among incommensurables—a difficulty that can be obscured but never wholly eliminated by any method of decision-making" [64]. An approach to this problem does not appear to be immediately forthcoming. Robert H. Socolow of Princeton University, in a discussion that applies remarkably well to Kaiparowits, notes that most major environmental decisions seem to get stuck in a rut:

> The discussions about whether to undertake substantial transformations of natural areas—to bring about new power plants, dams, airports, pipelines, deep water ports—have several pathologies in common. A cluster of detailed technical analyses accompanies the formulation of the program and its initial rush onto the stage; the proponents of the project imply, and generally believe, that all one could reasonably have expected has been done, both to justify the program and to anticipate its pitfalls. As after a carefully planned transplant, the reaction of rejection is slow in coming but grows relentlessly. The analyses are shown to be incomplete, and new analyses starting from different premises are eventually produced by those who wish to stop the program. But, contrary to what one might naively expect, the existence of disparate analyses does not help appreciably to resolve the debate. Rarely are the antagonists proud of their analyses; more rarely still are they moved by the analyses of their opponents. The combatants on both sides have been constrained by mandated rules of procedure as well as by the tactics of compromise. Understandably, the politicians in a position to determine the outcome conclude that their time is not well spent pondering the available analyses, even though they may commission still more of them. . . . Some environmental expert is presumed to be available who will come out with any answer for which a combatant is willing to pay. The analyst's results are presumed to be little more than the packaging of opinion and sentiment [65].

Part of the confusion stems from the fact that standard tools of analysis, based on an economic cost-benefit paradigm, are simply not suited for problems where there is little or no general agreement on what the costs and benefits are, or on the values that determine how the costs and benefits are to be compared and weighed. As Professors David Bradford and Harold Feiveson of Princeton note, "Political problems cannot be turned as if by magic into economic problems by waving a wand of benefit-cost analysis" [66]. In the case of Kaiparowits, how was one to measure the importance of natural beauty or ecological purity in the first place, let alone when the threat to these values was hard to assess? In a context of value controversy, analysis is somewhat limited. It can only clarify conflicts between values—it cannot resolve them.*

But the real problem that Kaiparowits points up is more social than analytical. The past two decades have marked a radically changing perception of the ways in which society and nature interact; only recently has society begun to realize that "nature modulates society," as Socolow puts it [69]. "The environment" became an object of popular concern; a formal discipline, ecology, was developed; an environmental movement was an inevitable occurrence. Such changes threw the hierarchy of social values based on economic growth out of alignment. People in general

---

* Policy analysis can fulfill creative as well as clarifying functions. As Henry S. Rowen remarks, ". . . for many participants the analytic process will contribute to beliefs about facts and relationships and will help in the construction of value preferences" [67]. Thus analysis can catalyze the modification of values. In addition, Rowen states, "the invention of new possibilities by policy analysis may help shift policy choices away from perhaps intractable zero-sum choices to nonzero-sum choices—from choices where what some people gain others lose to those where there are gains for all" [68]. By suggesting design changes in a project, for example, or by indicating new ways to pursue project goals, policy analysis can provide bases for consensus where only conflict was conceivable before.

do not know where to put their foot down regarding pollution, or how such "externalities" should be represented in the economic system. Development interests assure that technical fixes can solve supposed environmental problems, privately wondering what all the fuss is about in the first place. Environmental interests, meanwhile, play a delaying game for the most part, blocking developments but skirting the theoretical and practical issue of how to maintain a viable economy while protecting the environment to their satisfaction.

The fact that no decision on Kaiparowits was made by the federal authorities is perhaps representative of the present lack of a direction in response to the analytical and social problems sketched above. Conflicting interests have been institutionalized, but no consensus or authority exists for choosing between them. Utility officials feel stymied; environmentalists believe that they have made some inroads, but that their program is far from being implemented. Analysis and decision-making proceeds on a plant-by-plant basis, even though energy and environmental costs and benefits ultimately follow from energy development paths at large involving numbers, types, and locations of projects. However, even if a proposed project is rejected, there is no guarantee that the alternatives would be implemented.

Responsibility and authority are fragmented, but perhaps rightfully so, given conflicting values and difficulties in generating rational analyses of the trade-offs involved. The utility must try to meet demand without the support of a "proelectricity" consumer group or any other constituency. Each government agency has, to a certain extent, its own mandate; and the environmentalists have theirs. Secretary Kleppe's mandate, the priority of values that he would act upon, remained unclear. Even if Kleppe would have approved the project, the actual fate of Kaiparowits would have been left to the courts—a realm quite detached from the exigencies of planning electricity supply or developing reasoned environmental policies for the Southwest. In fact, at each stage of the game, as an actor gives a go-ahead to a project, he loses control over that project.

What Kaiparowits history ultimately portends is unclear. Kaiparowits may have had a negative effect on neighboring parklands during certain meteorological conditions, the extent dependent on the magnitude of other energy development in the vicinity. Environmentalists had successfully raised the possibility of a continuing, slow, serious deterioration of air quality in the region if a number of additional large coal-burning plants were constructed. The Arab oil embargo and the subsequent recession granted a temporary reprieve from growing electricity demand in the Kaiparowits market areas; demand has begun to climb back up, but had not reached its historic growth rate by 1977. In its planning, SCE is hoping to diversify its generation by utilizing combined cycle, coal, nuclear, and alternative (solar, coal gasification) technologies as needed, but it does not really know what the future holds.

Certainly the Kaiparowits case calls into question the prospects of extensive energy development in the Southwest or even an increasing commitment to traditional coal electricity-generating technologies foreseen in both the Project Independence and the Carter energy plans. The basic social conflict over the significance of given amounts of pollution and the determination of environmental values remains. These questions are more often than not sidestepped in the policy process and in the public sphere, where there is little sustained discourse about them. Moreover, it is unclear what the forum would be for a useful debate and effective social resolution of conflicts regarding the relationships between energy expansion, economic growth and the environment, or between conservation, the economy, and social well-being. Professor Socolow remarks that "[d]iscussions of goals, of visions of the future, are enormously inhibited. Privately, goals will be talked about readily . . . But the public debate is cloaked in a formality that excludes a large part of what people most care about" [70].

It is unsettling to note that Thomas Kleppe and James Schlesinger have stated independently that meaningful reform in energy matters will not occur until the lights go out. Many utility officials in the West concur, thinking that electricity curtailments could be required in the not-too-distant future. Environmentalists respond that a large-scale commitment to expanding the use of thermal generating technologies would likely lead to irreparable ecological damage, and that curtailments are by no means inevitable if a transition to nonconventional technologies and conservation is made—a pie-in-the-sky notion to most utility representatives. Whether this social conflict will work itself out before either spectre presents itself is a rather sobering question to ask.

APPENDIX A
The Clean Air Act

Congress enacted the Clean Air Act of 1970 (42 U.S.C. § 1857 *et. seq.* [1970]) in order to strengthen substantially the nation's existing policy against air pollution. The Act directed the Environmental Protection Agency (EPA) to formulate nationally uniform primary ambient air quality standards (designed to protect human health) and secondary ambient standards (designed to protect other aspects of human welfare) for specific air pollutants such as sulfur oxides, particulate matter, carbon monoxide, photochemical oxidants, hydrocarbons, and nitrogen dioxide.

Each state was required to develop a state implementation plan (SIP) to attain the primary ambient standards no later than May 10, 1977, and secondary standards within "a reasonable time" by imposing specific emission limitations on discharges from stationary sources and utilizing land use, transportation management, and other controls to reduce emissions from mobile sources. EPA was required to determine that the SIPs would in fact secure national ambient standards.

The Act also directed the EPA to establish national uniform emission standards requiring the adoption of best available emission control technology for newly constructed or modified major stationary pollution sources.

The Clean Air Act, coupled with the 1972 Water Pollution Control Act amendments, granted regulatory power to the EPA Administrator far beyond that possessed by other traditional federal regulatory commissions. These two environmental acts do not regulate specific industry types or practices but rather regulate nearly all industry and affect the behavior of all citizens through transportation and other controls.

Supreme Court action in 1973 led to a significant development in the history of the Act. In *Fri v. Sierra Club* (412 U.S. 541 [1975]), the court ruled that the EPA Administrator had a mandatory duty to insure that SIPs provide for nondegradation of air quality in those areas already meeting or exceeding primary and secondary standards. Approximately 80 per cent of the nation's lands have air quality that exceeds these standards. In response to this decision, the EPA Administrator in 1975 issued regulations requiring the states to divide their lands into three classes representing different allowable increases in total suspended particulates and sulfur dioxide. Class I applies to areas in which practically any change in air quality would be considered significant; Class II permits deterioration normally accompanying moderate, well-controlled growth; Class III applies to those areas in which deterioration up to the national standards would be considered insignificant.

All areas of the country were initially classified as Class II, but redesignation could be proposed by states, federal land managers, or Indian governing bodies. A federal land manager could choose to keep the air quality over federal lands in a more pristine condition than the state might designate. A pollution source cannot be sited if its operation would violate the air quality standards in the area where it would be located or the standards of any neighboring air quality area.

In response to the energy crisis, Congress amended the Clean Air Act in 1974 to permit one-year temporary exemptions from the 1970 primary and secondary emission standards for stationary emission sources (PL 93319). The rationale for this amendment was to facilitate the substitution of "dirty" but relatively plentiful high-sulfur coal for oil as fuel in power plants.

The Clean Air Act was amended again in 1977 (42 U.S.C. 7401) in order to reaffirm the policy of prevention of significant deterioration and to "preserve, protect, and enhance air quality in the national parks." All federal national monuments, national primitive areas, and national recreation areas exceeding 6000 acres in size (which would include almost all of the major national parks) were declared to be mandatory Class I areas and were no longer subject to discretionary state classification. The Amendment also set new quantitative standards for the maximum allowable increase in the particulate and sulfur dioxide levels for Class I areas. In addition, the Amendment declared a national goal of preventing any further impairment of visibility in mandatory Class I regions.

Congress mandated that no major emitting facility on which construction is commenced after the date of the Amendment's enactment (August 7, 1977) could be constructed unless it receives the appropriate permits, conforms to applicable emission standards, utilizes "best available control technology," and is appropriately monitored. A project is under construction if the owner has obtained all the necessary preconstruction approvals and either has begun a continuous construction program or has entered into binding monetary agreements to undertake a construction program that can be completed within a "reasonable time," agreements that cannot be cancelled without a substantial loss. If Kaiparowits would have produced pollution levels in the nearby parks in excess of Class I standards, it would have had to have been under construction in the above sense by August 7, 1977, to have been allowed.

APPENDIX B
Emission Rates

	Emission[a]					
	Average Grade Coal			Worst Grade Coal		
	Particulate	SO_2	NO_x	Particulate	SO_2	NO_x
Tons per hour[b]	0.43	1.15	10.42	0.58	2.17	10.42
Tons per year[c]	2,820	7,555	68,000			

[a] 90 percent control of SO_2; 99.5 percent control of particulates; and 32.5 percent control of NO_x.
[b] 100 percent load factor.
[c] 75 percent load factor.

Estimated Full-Load[a] Emissions and Emission Rates Resulting from Regulation Compliance and Proposed Control

Pollutants	NSPS[b] Emission Limitations (lb/10^6 Btu)	Emission Rates (ton/hour)	Proposed Emission Control (percent)	Proposed Control Emission Rate (lb/10^6 Btu)		Proposed Control Emission Rate (tons/hour)	
				Avg. Coal	Worst Coal	Avg. Coal	Worst Coal
Sulfur Dioxide	1.20	17.8	90.0	0.08	0.15	1.15	2.17
Particulate Matter	0.10	1.5[c]	99.5	0.03	0.04	0.43	0.58
Nitrogen Oxides	0.70	10.4	32.5	0.70	0.70	10.40	10.40

[a] Full load—29,680 Btu/hour.
[b] NSPS: federal New Source Performance Standards.
[c] The water contract with the Department of the Interior requires a design efficiency of 99.5 percent removal of particulate matter (0.43 and 0.58 ton/hour) with monthly averages not less than 97 percent (2.6 and 3.5 ton/hour) and a 24-hour average not less than 96 percent (3.5 and 4.7 ton/hour).

APPENDIX C
Applicable Emission Standards

Estimated Full-Load Emission Data for the Kaiparowits Plant[a] and Comparison with Federal Regulations for New Sources

Emission	Estimated Full-Load Emissions ($lb/Btu \times 10^6$) Average Grade Coal	Worst Grade Coal	Federal Regulations for New Plants
Sulfur Dioxide	0.08	0.15	1.20
Particulate Matter	0.03	0.04	0.10[b]
Nitrogen Oxides	0.70	0.70	0.70

[a] Based on 90 percent removal of SO_2, 99.5 percent particulate removal.

[b] The water contract with the Department of Interior requires a design efficiency of 99.5 percent removal of particulate matter with a monthly average not less than 97 percent and a 24-hour average not less than 96 percent.

Predicted Emissions and Applicable Emission Control Standards for Proposed Generating Plant
(3,000-MW, 29,680 $\times$ 10^6 Btu/hr at 1 Percent Load)

Emission	Without Emission Control (ton/hr)		Applicable Standard and Allowable Emission Rate (ton/hr)		Abatement Required To Meet Standard (%)		Committed Abatement by Applicant (%)	Predicted Abated Emission (ton/hr)	
	Average[a] Coal	Worst[b] Coal			Average[a] Coal	Worst[b] Coal		Average[a] Coal	Worst[b] Coal
SO_2	11.5	21.6	EPA	(17.8)	0.0	17.6	90	1.15	2.17
			Arizona	(11.9)	0.0	44.9			
			Utah[c]	(17.8)	0.0	17.6			
NO_x	15.4	15.4	EPA	(10.4)	32.5	32.5	32.5	10.40	10.40
			Arizona	(10.4)	32.5	32.5			
			Utah[c,d]						
Particulate	86.6	116.8	EPA	(1.48)	98.3	98.8	99.5	0.43	0.58
			Arizona	(0.73)	99.2	99.4			
			Utah[e]						
			Dept. of Interior[f]		99.5	99.5			

[a] Sulfur, 0.42% by weight; ash content, 7.00%; heating value, 10,800 Btu/lb; amount burned, 33,000 ton/day.
[b] Sulfur, 0.75% by weight; ash content, 9.00%; heating value, 10,300 Btu/lb; amount burned, 34,600 ton/day.
[c] Air pollution control equipment and processes shall be selected and operated to afford highest efficiencies and lowest discharge rates that are reasonable and practicable. New installations shall control sulfur dioxide emissions as required to avoid exceeding national primary and secondary ambient air quality standards and federal standards of performance for new stationary sources.
[d] No nitrogen oxide emission rate specified.
[e] No particulate emission rate for new sources specified under present regulation.
[f] Water service contract specifies a design efficiency for removal of 99.5 percent of particulate matter with a monthly average not less than 97 percent and a 24-hour average not less than 96 percent.

REFERENCES

[1] Alvin M. Josephy, Jr., "Kaiparowits, The Ultimate Obscenity," *Audubon Magazine*, March 1976 [hereinafter cited as *Josephy*].

[2] Bureau of Land Management, United States Department of Interior, *Final Environmental Impact Statement, Proposed Kaiparowits Project*, Chapter III, p. 40 (1976) [hereinafter cited as *Final EIS*].

[3] *Final EIS*, Ch. V, p. 22.

[4] *Final EIS*, Ch. III, pp. 261-262.

[5] *Final EIS*, Ch. V, p. 25.

[6] *Final EIS*, Ch. VII, p. 4.

[7] *Final EIS*, Summary.

[8] *Final EIS*, Ch. VI, p. 8.

[10] Southern California Edison, San Diego Gas and Electric, Arizona Public Service, *Participants' Position Statements Supporting Recommended Federal Decisions*, Kaiparowits Project, February 1976, pp. 8-11 [hereinafter cited as *Participants' Position Statement*].

[11] *Josephy*, p. 73.

[12] Environmental Defense Fund, *Comments on Environmental Impact Statement, Proposed Kaiparowits Project,* (no date), p. 6.

[13] Friends of the Earth, Inc., letter to Senator Henry M. Jackson, February 18, 1976, p. 4 [hereinafter cited as *FOE February 18*].

[14] *Josephy*, p. 66.

[15] Friends of the Earth, Inc., letter to Paul Howard, State Director, Bureau of Land Management, Utah, November 13, 1975, p. 15 [hereinafter cited as *FOE November 13*].

[16] *Id.*, p. 14.

[17] *Id.*

[18] United States Environmental Protection Agency, letter from Administrator Russell E. Train to Secretary of the Interior Thomas S. Kleppe, June 4, 1976, p. 5 [hereinafter cited as *EPA letter*].

[19] *Id.*

[20] *FOE February 18*, p. 4.

[21] National Park Service, United States Department of the Interior, *Analysis of Kaiparowits Powerplant Impacts on National Recreation Resources*, March 1970, p. 29 [hereinafter cited as *NPS Study*].

[22] *Id.*, pp. 31-34.

[23] *Id.*, p. 28.

[24] *Id.*, p. 17.

[25] *Final EIS*, pp. 261-262.

[26] *NPS Study*, p. 37.

[27] *Id.*, p. 3.

[28] *Id.*, p. 2.

[29] *Id.*, p. 11.

[30] *Final EIS*, Ch. III, p. 10.

[31] *Id.*, Ch. III, p. 64.

[32] *Id.*, Ch. III, p. 200.

[33] *Participants' Position Statement*, pp. 12-13.

[34] *EPA letter*, pp. 1-4.

[35] *Participants' Position Statement*, p. 5.

[36] *Final EIS*, Reference Material, p. A-115.

[37] *Id.*

[38] *EPA letter*, pp. 7-14.

[39] *FOE November 13*, pp. 7-8.

[40] United States Department of the Interior, Final Kaiparowits Program Decision Option Document (PDOD), April 30, 1976, p. 20 [hereinafter cited as *PDOD*].

[41] *FOE, November 13*, p. 25.

[42] *Id.*, p. 14.

[43] Department of the Interior news release, cited in *FOE, February 10*, p. 4.

[44] Information in this paragraph is from correspondence with Christian Hill, staff reporter, *The Wall Street Journal*, July 1977.

[45] *Id.*

[46] *Id.*

[47] L. T. Papay, "The Kaiparowits Story," Presentation at the American Chemical Society Symposium: Coal Dilemma, How and Where to Use It, February 2, 1977, pp. 12-13.

[48] The Senate Report on NEPA, S. Rep. No. 91-296, 91st Congress, 1st Session (1969), cited in Richard S. Mallory, *The Legally Required Contents of a NEPA Environmental Impact Statement*, Stanford Environmental Law Society, 1976, p. 48 [hereinafter cited as *Mallory*].

[49] NEPA, cited in *Mallory*, p. 2.

[50] *Id.*, p. 55.

[51] U. S. Department of the Interior, *Southwest Energy Study*, Summary Report, pp. FC-1, 2.

[52] Hanna J. Cortner, *A Case Analysis of Policy Implementation: The National Environmental Policy Act of 1969*, Lake Powell Research Project Bulletin, Number 10, May 1975, p. 2 [hereinafter cited as *Cortner*].

[53] *Id.*, p. 3.

[54] *Id.*, p. 14.

[55] Eugene Bardach and Lucian Pugliaresi, "The Environmental-Impact Statement vs. The Real World," *The Public Interest*, Fall 1977, p. 24 [hereinafter cited as *Bardach and Pugliaresi*].

[56] *Id.*, p. 27.

[57] *Id.*, p. 29.

[58] *Cortner*, p. 10.

[59] Scott C. Whitney, "The Case for Creating a Special Environmental Court System," cited in *Cortner*, p. 9.

[60] Werner F. Grunbaum, "Judicial Attitudes Toward Environmental Quality in the Federal and State Courts," unpublished paper presented at the Annual Meeting of the American Political Science Association, 1974, cited in *Cortner*, p. 14.

[61] *Bardach and Pugliaresi*, p. 36.

[62] *Id.*, p. 37.

[63] Much of the information in the following paragraphs was provided by Secretary Kleppe and other DOI officials during personal interviews.

[64] Lawrence Tribe, "Ways Not to Think About Plastic Trees," in *When Values Conflict, Essays on Environmental Analysis, Discounts, and Decision*, edited by Lawrence H. Tribe, Corinne S. Schelling, and John Ness, p. 65 [hereinafter cited as *Tribe et al.*]

[65] Robert H. Socolow, "Failures of Discourse," in *Tribe et al.*, pp. 1-2, 25.

[66] David Bradford and Harold Feiveson, "Benefits and Costs, Winners and Losers," in *Boundaries of Analysis, An Inquiry Into the Tocks Island Dam Controversy*, edited by Harold A. Feiveson, Frank W. Sinden, and Robert H. Cocolow, p. 157.

[67] Henry S. Rowen, "Policy Analysis as Heuristic Aid," in *Tribe et al.*, p. 142.

[68] *Id.*, p. 144.

[69] Socolow in *Tribe et al.*, p. 12.

[70] *Id.*

INDEX

L